Cosmological Redshift of Light

Cosmological Redshift of Light

by

Trevor G. Underwood

By the same author:

"Quantum Electrodynamics – annotated sources. Volumes I and II." (April 2023);

"Special Relativity." (June 2023);

"General Relativity." (November 2023);

"Gravity." (March 2024);

"Electricity & Magnetism." (May 2024);

"Quantum Entanglement." (June 2024);

"The Standard Model." (September 2024);

"New Physics." (October 2024).

Published by Trevor G. Underwood
18 SE 10th Ave.
Fort Lauderdale, FL 33301

ISBN: 979-8-218-55145-2 (hardcover)

Library of Congress Control Number: 2024924033

Printed and distributed by Lulu Press, Inc.
627 Davis Dr.
Ste. 300
Morrisville, NC 27560
http://www.lulu.com/shop

Front cover: A picture of the inner regions of the famous Crab Nebula captures emergent jets and the "Napoleon Hat" structure of surrounding plasma. The radio/optical/X-ray pulsar, a neutron star rotating at ~30 Hz, is buried in the center. The Crab was produced in a supernova explosion in A.D. 1054. Image courtesy of ESA/NASA. [Burrows. A. S. (February, 2015). Baade and Zwicky: "Super-novae," neutron stars, and cosmic rays. *Proceedings of the National Academy of Sciences*, 112, 5, 1241, Fig. 1.]

CONTENTS.

Page no.

3

Compton's paper noted that classical electrodynamics predicts that the energy scattered by an *electron* traversed by an *X-ray* beam is independent of the *wave-length* of the incident rays. It also predicts that when the *X-rays* traverse a thin layer of matter the intensity of the scattered radiation on the two sides of the layer should be the same. But experiments on the scattering of *X-rays* by light elements showed that these predictions were correct when *X-rays* of moderate hardness were employed, but when very hard *X-rays* or *γ-rays* were employed, the scattered energy was less than Thomson's theoretical value and was strongly concentrated on the emergent side of the scattering plate. Compton applied Einstein's hypothesis to the scattering of *X-ray* and *γ-ray photons* by *electrons* and derived the mathematical relationship between the shift in *wavelength* and the scattering angle of the *X-rays, by assuming that each scattered X-ray photon interacts with only one electron.* This agreed with experimental results for the scattering of *X-ray* and *γ-ray photons* by *electrons,* subsequently known as *Compton scattering,* providing important evidence for *quantum theory.* The introduction of *special relativity* was irrelevant to the comparison of the theory with experimental results.

82 **Arthur Holly Compton (September 10, 1892 – March 15, 1962).**

89 **Baade, W. (1938). The Absolute Photographic Magnitude of Supernovae.** *Astrophys. J.,* 88, 285-304; https://articles.adsabs. harvard.edu/pdf/1938ApJ....88..285B; a compilation of the *photometric data* for the 18 *supernovae* known at the end of 1937 is given. Former estimates have been replaced by *photometric magnitudes* after a redetermination of the magnitudes of comparison stars on the international system. The mean absolute *photographic magnitude* of the *supernovae,* derived from this material, is $M^-_{max} = -14.3 \pm 0.42$ (m. e.) with a dispersion $M_{max} \approx 1.1$ mag. This result, together with the spectroscopic evidence, fully confirms the view that two classes of *novae, common novae* and *supernovae,* exist. Attention is drawn to the curious fact that 72 per cent of the known *supernovae* appeared in late-type spirals. *B Cassiopeiae* and the *Crab nebula,* which may have been galactic *supernovae,* are discussed.

93 **Burrows. A. S. (February, 2015). Baade and Zwicky: "Super-novae," neutron stars, and cosmic rays.** *Proc. Natl. Acad. Sci.,* 112, 5, 1241-2; https://www.pnas.org/doi/epdf/10.1073/pnas. 1422666112; in 1934, two astronomers in two of the most prescient papers in the astronomical literature [Baade, W., & Zwicky, F. (1934). On super-novae. *Proc. Natl. Acad. Sci.,* 20, 5,

254–9; Baade, W., & Zwicky, F. (1934). Cosmic rays from super-novae. *Proc. Natl. Acad. Sci.*, 20, 5, 259–63] coined the term "*supernova*", hypothesized the existence of *neutron stars*, and knit them together with the origin of *cosmic-rays* to inaugurate one of the most surprising syntheses in the annals of science.

99 **Tonry, J. & Schneider, D. P. (September, 1988). A New Technique for Measuring Extragalactic Distances.** *Astron. J.*, 96, 3, 807; https://articles.adsabs.harvard.edu/pdf/ 1988AJ.....96..807T; this paper describes a relatively direct technique of determining *extragalactic distances*. The method relies on measuring the *luminosity fluctuations* that arise from the *counting statistics* of the *stars* contributing the *flux* in each pixel of a high-signal-to-noise CCD (*Charge Coupled Device*) image of a *galaxy*. *The amplitude of these fluctuations is inversely proportional to the distance of the galaxy.* This approach bypasses most of the successive stages of calibration required in the traditional *extragalactic distance ladder*; the only serious drawback to this method is that it requires an accurate knowledge of the bright end ($M_v < 3$) of the *luminosity function*. Potentially, this method can produce accurate distances of *elliptical galaxies* and *spiral bulges* at distances out to about 20 Mpc. The paper explains how to calculate the value of the fluctuations, taking into account various sources of contamination and the effects of finite spatial resolution, and demonstrates, via simulations and CCD images of M32 and N3379, the feasibility and limitations of this technique.

104 **The expanding universe and the Big Bang.**

109 **Perlmutter, S., Aldering, G., Goldhaber, G., Knop, R.A., Nugent, P.,** *et al.* **(December, 1998). Measurements of Omega and Lambda from 42 High-Redshift Supernovae.** *Astrophys. J.*, 517, 565-86; https://arxiv.org/pdf/astro-ph/9812133; this paper reports on the *Supernova Cosmology Project*, which was

started in 1988 to determine the cosmological parameters of the universe using the *magnitude-redshift* relation of Type Ia *supernovae*. All *supernova* peak magnitudes are standardized using a SN Ia *light-curve width-luminosity relation*. It determines that the data are *strongly inconsistent* with a $\Lambda = 0$ *flat cosmology*, the simplest inflationary universe model in which the universe continues to expand at a *constant rate*, for which the best-fit age of the universe relative to the Hubble time was $t_0^{flat} = 14.9_{-1.1}^{+1.4}$ (0.63/h) Gyr. The data indicate that the *cosmological constant* is *non-zero* and *positive*, indicating an *accelerating universe*. This relation was based on a *relativistic cosmological model*, so is inconsistent with New Physics.

120 **Lubin, L. M. & Sandage, A. (June, 2001). The Tolman Surface Brightness Test for the Reality of the Expansion. IV. A Measurement of the Tolman Signal and the Luminosity Evolution of Early-Type Galaxies.** *Astrophys. J.*, 517, 565-86; https://arxiv.org/pdf/astro-ph/0106566; this article was included because it was one of very few recent "mainstream" articles which claimed to evaluate *Zwicky's "tired-light" theory*, but in fact it did not. Although Allan Sandage was one of the most influential astronomers of the 20th century, he was 75 years old at this time. It was very disappointing. It omitted an expansion factor, then failed to include the *"tired-light" factor* representing the linear loss of energy as the *photons* interact with *electrons* and other matter as they travel through intergalactic space. The evidence for an *expanding universe* which it provided based on *Tolman surface brightness test* was demonstrated by Lerner, E. J., Falomo, R., & Scarpa, R. (2014). *UV surface brightness of galaxies from the local universe to z ~ 5*, to be in error, and, when calculated correctly, *concluded that far from disproving a non-expanding cosmology, data by Lubin and Sandage agreed very well with predictions for a static Euclidean universe.* However, this paper provides some useful insights on this analysis, and on the deeply ingrained biases at work.

147 **Seife, C. (June, 2001). 'Tired-Light' Hypothesis Gets Re-Tired.** *Science*, 292, 5526, 2414; https://www.science.org/doi/10.1126/science.292.5526.2414a; article by journalist attacking the "tired-light" hypothesis based on the recently published articles (1) Perlmutter, S., Aldering, G., Goldhaber, G., Knop, R.A., Nugent, P., *et al.* (December, 1998). Measurements of Omega and Lambda from 42 High-Redshift Supernovae, and (2) Lubin, L. M. & Sandage, A. (June, 2001). The Tolman Surface Brightness Test for the Reality of the Expansion. IV. A Measurement of the Tolman Signal and the Luminosity Evolution of Early-Type Galaxies (see above). Both depend on *relativistic* assumptions, which do not apply in New Science.

150 **Mamas, D. L. (2010). An explanation for the cosmological redshift.** *Phys. Essays*, 23, 326; https://tiredlight.net/wp-content/uploads/2014/09/physics-essays-2010.pdf; this is a bizarre attempt by an unrecognized author based in Florida (with a PhD in physics from UCLA) to reformulate *Zwicky's "tired-light" theory* in terms of a simplistic pseudo-classical notion of a *photon* as an electromagnetic wave which causes a *free electron* to oscillate and reradiate. It is included here as it provides useful background in the absence of any mainstream paper. It also provides calculations that suggest that the *cosmological redshift*, based on the quantum mechanical formulation of *Zwicky's "tired-light" theory* and estimates of the density of *free electrons* in intergalactic space, support a *static* rather than an *expanding universe*, on which the *Big Bang* origin of the universe is based. It incorrectly assumes, as Zwicky stated, that *Compton scattering* suffers from the blurring of images.

162 **The effect of a strong magnetic field on a dielectric.**

164 **Underwood, T. G. (October 12, 2024). The Faraday effect.**

172   **Shaoa, M-H. (April, 2013). The energy loss of photons and cosmological redshift.** Published online 8 April 8, 2013; http://dx.doi.org/10.4006/0836-1398-26.2.183; this is another rather bizarre article in which Zwicky's *"tired-light"* theory is presented in terms of a simplistic interpretation of the wave-particle nature of the *photon* and classical electromagnetic theory. This is rather surprising as it was published by Xinjiang Astronomical Observatory, Key Laboratory of Radio Astronomy, People's Republic of China, though not by a mainstream theoretical physics institution. It is included for the same reasons as Mamas, D. L. (2010).

188   **Lerner, E. J., Falomo, R., & Scarpa, R. (2014). UV surface brightness of galaxies from the local universe to z ~ 5.** *International Journal of Modern Physics* D, 23, 06, 1450058; https://doi.org/10.1142/S0218271814500588; also at https://arxiv.org/pdf/1405.0275; an important article that demonstrates errors in previous claims that have been made that the Tolman test provides compelling evidence against a static model for the universe. This was reconsidered by adopting a *static Euclidean universe* (SEU) with *a linear Hubble relation at all z* (which is not the *standard Einstein–de Sitter model*), resulting in a relation between *flux* and *luminosity* that is virtually indistinguishable from the one used for ΛCDM models. Based on the analysis of the UV SB of luminous disk galaxies from HUDF and GALEX datasets, reaching from the local universe to z ~ 5, it was shown that the *surface brightness* (SB) *remains constant as expected in a static universe.*

In particular, it shows that the conclusions in Lubin & Sandage *are not supported by the data* for two main reasons: (1) for the *static scenario*, Lubin and Sandage set the distance to d = (c/H$_0$) ln(1 + z), *which is valid only for the Einstein-de Sitter static case.* This is not the cosmology being tested where the *Hubble relation* is hypothesized to be d = cz/H$_0$ at all *redshift*. The conversion factors to transform arc seconds to pc in the *non-expanding model* are therefore different; (2) *the local sample includes only first rank cluster galaxies, while the high-z sample includes about 20 normal galaxies in each of three different clusters.* This means that their distant galaxies are on average smaller and less luminous, and therefore are not directly comparable to local ones because of the well-known *absolute magnitude-SB relation. It concludes that far from disproving a non-expanding cosmology, data by Lubin and Sandage agree very well with predictions for a static Euclidean universe.*

210 **Mamas, D. L. (May, 2015). Cosmological redshift model now experimentally confirmed.** *Physics Essays*, 28, 2, 201-2; http://dx.doi.org/10.4006/0836-1398-28.2.201; a theoretical model [Mamas, D. L. (2010). *Phys. Essays*, 23, 326], which accounts for the *cosmological redshift* in a static universe, now has experimental confirmation. In this model, the *photon* is viewed as an *electromagnetic wave* whose *electric field* component causes oscillations in deep space *free electrons which then reradiate energy from the photon,* causing a *redshift.* Calculations from the model match well the anomalous *redshift* of Wernher von Braun's Pioneer 6 spacecraft.

PREFACE

This book addresses the cause of the *cosmological redshift of light*. It is currently assumed to be due to the *Doppler effect* on light resulting from the *expansion of the universe* following the *Big Bang*. However, there is an alternative less radical theory. Fritz Zwicky's *"tired-light" theory*, attributes the *linear redshift with distance from the observer* to the *loss of energy by photons*, and consequent increase in *wavelength*, resulting from *interactions between the photons and intervening electrons or matter* whilst travelling through *intergalactic* space.

It is possible that both phenomena contribute to the observed redshift but if the *loss of energy by photons* is sufficient, this would suggest that *there was no Big Bang, the universe is not just 13.8 billion years old, but is indefinitely old, and in a steady state, not expanding*. In the absence of discussion by recognized institutions, the analysis below, that has been pieced together from disparate, largely non-mainstream, sources, suggests that is the case.

The notion of an *expanding universe* was first scientifically originated by physicist Alexander Friedmann in 1922, a Russian cosmologist and mathematician, who derived what became known as the Friedmann equations from the field equations of *Einstein's theory of general relativity*, showing that the universe might be expanding in contrast to the static universe model advocated by Albert Einstein at that time.

Independently deriving Friedmann's *relativistic* equations in 1927, Georges Lemaître, a Belgian physicist and Roman Catholic priest, proposed that the presumed recession of the nebulae was due to the *expansion of the universe*. He inferred the relation between the *redshift* and *distance* that Hubble would later observe. In 1931, Lemaître went further and suggested that the evident *expansion of the universe*, if projected back in time, meant that the further in the past the smaller the universe was, until at some finite time in the past all the mass of the universe was concentrated into a single point, a "primeval atom" where and when the fabric of time and space came into existence, in what was later referred to as the *"Big Bang singularity"*.

In the 1920s and 1930s, almost every major cosmologist preferred an eternal *steady-state universe*. During the 1930s, other ideas were proposed as non-standard cosmologies to explain Hubble's observations, including Fritz Zwicky's *"tired-light" hypothesis*.

Subsequently, various cosmological models of the *Big Bang* attempted to explain the evolution of the observable universe from the earliest known periods through its subsequent large-scale form. They offered an explanation for a broad range of observed phenomena, including the abundance of light elements, the *cosmic microwave background* (CMB) radiation, and large-scale structure. The models depended on two major assumptions: the *universality of physical laws* (one of the underlying principles of the *theory of relativity*) and the *cosmological principle*. These models were compatible with the Hubble–Lemaître law—the observation that *the farther away a galaxy is, the faster it appeared to be moving away from Earth*. Under this theory, detailed measurements of the *redshift* placed the *Big Bang singularity* at an estimated 13.8 billion years ago, which was considered the *age of the universe*.

In 1929, Edwin Hubble published his famous paper in which he used the strong direct relationship between a classical Cepheid variable's *luminosity* and *pulsation period* for scaling galactic and extragalactic *distances* and confirmed that the *redshift* of a galaxy, expressed as its *radial velocity*, increases with its *distance* from Earth, a behavior that became known as Hubble's law. [Hubble, E. (April, 1929). *A relation between distance and radial velocity among extra-galactic nebulae*, below.]

In response to Hubble's announcement of this somewhat linear relationship, Fritz Zwicky immediately pointed out that the correlation between the calculated *distances* of galaxies and their *redshifts* had a discrepancy too large to fit in the *distance's* error margins. [Zwicky, F. (1929). *On the Red Shift of Spectral Lines through Interstellar Space*, below.] Zwicky *proposed his* "tired-light" theory that *the reddening effect was not due to motions of the galaxy, but to an unknown phenomenon that caused photons to lose energy as they traveled through space.* He considered the most likely candidate process to be a drag effect in which *photons* transfer *momentum* to surrounding masses through *gravitational interactions.* [However, *photons* have no *mass.*] He rejected explanations involving the *expansion of space,* and incorrectly stated that *Compton scattering* suffers from the blurring of images.

The MIT OpenCourseWare article below, "*Interactions of Photons with Matter*", describes three *energy loss mechanisms,* the *photoelectric effect; Compton scattering;* and *pair production.* It notes that because photons are electrically neutral, they *do not steadily lose energy* via coulombic interactions with atomic *electrons,* as do charged particles. Photons travel some considerable distance before undergoing a more "*catastrophic*" interaction leading to *partial or total transfer* of the *photon* energy to *electron* energy. These *electrons* will ultimately deposit their energy in the medium. Photons are far *more penetrating* than *charged particles* of similar *energy.*

The *photoelectric* process is the *predominant mode of photon interaction* at *relatively low photon energies and high atomic number.* In the *photoelectric absorption process,* a *photon* undergoes an interaction with an *absorber atom* in which the *photon* completely disappears. In its place, an energetic *photoelectron* is ejected from one of the bound shells of the *atom.* For *gamma rays* of sufficient energy, the most probable origin of the *photoelectron* is the most tightly bound or K shell of the atom. For *gamma-ray* energies of more than a few hundred keV, the *photoelectron* carries off the majority of the original *photon energy.* The *photoelectric interaction* is most likely to occur if the energy of the incident *photon is just greater than the binding energy* of the *electron* with which it interacts.

Compton scattering, named after Arthur Compton an American physicist who won the Nobel Prize in Physics in 1927 for his 1923 discovery of the Compton effect, takes place between the incident *gamma-ray photon* and an *electron* in the absorbing material. In Compton scattering, the incoming *gamma-ray photon* is *deflected* through an angle θ with respect to its original direction. The *photon* transfers a portion of its *energy* to the *electron* (assumed to be initially at rest), which is then known as a *recoil electron*, or a *Compton electron*. All angles of scattering are possible. The *energy* transferred to the *electron* can vary from zero to a large fraction of the *gamma-ray energy*. The *Compton scattering* probability is almost *independent of atomic number*; decreases as the *photon energy* increases; is *directly proportional* to the number of *electrons* per gram, which only varies by 20% from the lightest to the heaviest elements (except for hydrogen).

If a *photon* enters matter with an energy *in excess of 1.022 MeV*, it may interact by a process called *pair production*. The *photon*, passing near the *nucleus* of an atom, is subjected to strong field effects from the *nucleus* and may disappear as a *photon* and reappear as a positive and negative *electron pair*. The two *electrons* produced, e⁻ and e⁺, are not scattered *orbital electrons*, but are created, *de novo*, in the *energy/mass conversion* of the disappearing *photon*. The *kinetic energy* of the *electrons* produced will be the difference between the *energy* of the incoming *photon* and the energy equivalent of two *electron* masses (2 x 0.511, or 1.022 MeV). *Pair production probability*, increases with *increasing photon energy*; and increases approximately as the square of the *atomic number*.

The *bulk behavior of photons in an absorber* is given by the *probability of an interaction per unit distance traveled*, $N = N_0 e^{-\mu x}$, where the linear *attenuation coefficient* μ, has the dimensions of inverse length (eg. cm⁻¹) and depends on *photon energy* and on the *material being traversed*; and the *mass attenuation coefficient*, μ/ρ, is obtained by dividing μ by the *density* ρ of the material, usually expressed in cm²g⁻¹.

Compton's paper [Compton, A. H. (May, 1923). *A Quantum Theory of the Scattering of X-rays by Light Elements*] confirmed that the scattering of *electrons* by *X-rays* or *γ-rays* results in a *redshift* due to the transfer of *energy* from the *photons* to the scattered *electron*. For *electromagnetic radiation* from remote *galaxies* observed from the Earth travelling through *intergalactic* space, known to contain *electrons* and other matter, this results in a *linear increase in the redshift with distance travelled by the photons*. This contributes to the *cosmological redshift of light*, which is currently assumed to be due to the *Doppler effect* on light resulting from the *expansion of the universe* following the *Big Bang*. It is possible that both phenomena contribute to the observed *redshift* but if the *loss of energy by photons* is sufficient, this would suggest that *there was no Big Bang, the universe is not just 13.8 billion years old*, but *is indefinitely old, and in a steady state, not expanding.*

Compton's paper noted that classical electrodynamics predicts that the *energy* scattered by an *electron* traversed by an *X-ray* beam is independent of the *wave-length* of the incident rays. It also predicts that when the *X-rays* traverse a thin layer of *matter* the *intensity* of the scattered radiation on the two sides of the layer should be the same. But experiments on the scattering of *X-rays* by light elements showed that these predictions were correct when *X-rays* of moderate hardness were employed, but when very hard *X-rays* or *γ-rays* were employed, the scattered energy was less than Thomson's theoretical value and was strongly concentrated on the emergent side of the scattering plate. Compton applied Einstein's hypothesis to the scattering of *X-ray* and *γ-ray photons* by *electrons* and derived the mathematical relationship between the shift in *wavelength* and the scattering angle of the *X-rays by assuming that each scattered X-ray photon interacts with only one electron*. This agreed with experimental results for the scattering of *X-ray* and *γ-ray photons* by *electrons*, subsequently known as *Compton scattering*, providing important evidence for *quantum theory*. [The introduction of *special relativity* was irrelevant to the comparison of the theory with experimental results.]

Burrows. A. S. (February, 2015). *Baade and Zwicky: "Super-novae," neutron stars, and cosmic rays*, describes how, in 1934, two astronomers in two of the most prescient papers in the astronomical literature coined the term *"supernova"*, hypothesized the existence of *neutron stars*, and knit them together with the origin of *cosmic-rays* to inaugurate one of the most surprising syntheses in the annals of science. [Baade, W., & Zwicky, F. (1934). On super-novae. *Proc. Natl. Acad. Sci.*, 20, 5, 254–9; Baade, W., & Zwicky, F. (1934). Cosmic rays from super-novae. *Proc. Natl. Acad. Sci.*, 20, 5, 259–63].

In Baade, W. (1938). *The Absolute Photographic Magnitude of Supernovae*, Walter Baade provided a compilation of the *photometric data* for the 18 *supernovae* known at the end of 1937. Former estimates were replaced by *photometric magnitudes* after a redetermination of the magnitudes of comparison stars on the international system. The mean absolute *photographic magnitude* of the *supernovae*, derived from this material, was $M^-_{max} = -14.3 \pm 0.42$ (m. e.) with a dispersion $M_{max} \approx 1.1$ mag. This result, together with the spectroscopic evidence, fully confirmed the view that two classes of *novae, common novae* and *supernovae*, exist. Attention was drawn to the curious fact that 72 per cent of the known *supernovae* appeared in late-type spirals. *B Cassiopeiae* and the *Crab nebula*, which may have been galactic *supernovae*, were discussed.

Tonry, J. & Schneider, D. P. (September, 1988). *A New Technique for Measuring Extragalactic Distances*, describes a relatively direct technique of determining *extragalactic distances*. The method relies on measuring the *luminosity fluctuations* that arise from the *counting statistics* of the *stars* contributing the *flux* in each pixel of a high-signal-to-noise CCD (*Charge Coupled Device*) image of a *galaxy. The amplitude of these fluctuations is inversely proportional to the distance of the galaxy.* This approach bypasses most of the successive stages of calibration required in the traditional *extragalactic distance ladder*; the only serious drawback to this method is that it requires an accurate knowledge of the bright end ($M_v < 3$) of the *luminosity function*. Potentially, this method can produce accurate distances of *elliptical galaxies* and *spiral bulges* at distances out to about 20 Mpc. The paper

explains how to calculate the value of the fluctuations, taking into account various sources of contamination and the effects of finite spatial resolution, and demonstrates, via simulations and CCD images of M32 and N3379, the feasibility and limitations of this technique.

In *"The expanding universe and the Big Bang"*, the development of these theories is described. The notion of an *expanding universe* was first scientifically originated by physicist Alexander Friedmann in 1922 with the mathematical derivation of the Friedmann equations from the field equations of *Einstein's theory of general relativity*. Independently deriving Friedmann's *relativistic* equations in 1927, Georges Lemaître, a Belgian physicist and Roman Catholic priest, proposed that the recession of the *nebulae* was due to the *expansion of the universe*.

In 1931, Lemaître went further and suggested that the evident *expansion of the universe*, if projected back in time, meant that the further in the past the smaller the universe was, until at some finite time in the past all the mass of the universe was concentrated into a single point, a "primeval atom" where and when the fabric of time and space came into existence.

Fred Hoyle coined the phrase that came to be applied to Lemaître's theory, referring to it as "this *Big Bang* idea" during a BBC Radio broadcast in March 1949. The *Big Bang* is a physical theory that describes how the universe expanded from an initial state of high density and temperature.

Various cosmological models of the *Big Bang* attempted to explain the evolution of the observable universe from the earliest known periods through its subsequent large-scale form. They offered an explanation for a broad range of observed phenomena, including the abundance of light elements, the *cosmic microwave background* (CMB) radiation, and large-scale structure. Detailed measurements of the expansion rate of the universe placed the *Big Bang* singularity at an estimated 13.8 billion years ago, which was considered to be the *age of the universe*.

Perlmutter, S., Aldering, G., Goldhaber, G., *et al.* (December, 1998). *Measurements of Omega and Lambda from 42 High-Redshift Supernovae*, reports on the *Supernova Cosmology Project*, which was started in 1988 to determine the cosmological parameters of the universe using the *magnitude-redshift* relation of Type Ia *supernovae*. All *supernova* peak magnitudes were standardized using a SN Ia *light-curve width-luminosity relation*. It determined that the data are *strongly inconsistent* with a $\Lambda = 0$ *flat cosmology*, the simplest inflationary universe model in which the universe continues to expand at a *constant rate*, for which the best-fit age of the universe relative to the Hubble time was $t_0^{flat} = 14.9_{-1.1}^{+1.4}$ (0.63/h) Gyr. The data indicated that the *cosmological constant* was *non-zero* and *positive*, indicating an *accelerating universe.* This relation was based on a *relativistic cosmological model*, so is inconsistent with New Physics*.

[* My previous books, *"Quantum Electrodynamics – annotated sources. Volumes I and II."* (April 2023), *"Special Relativity."* (June 2023), *"General Relativity."* (November 2023), *"Gravity."* (March 2024), *"Electricity & Magnetism."* (May 2024), *"Quantum Entanglement."* (June 2024), *"The Standard Model."* (September 2024), and "New Physics." (October 2024), came to the conclusion that the Standard Model of Particle Physics needs to be replaced with a far less complicated *non-relativistic* New Physics.

New Physics is a formulation of physics in which *Einstein's theory of Special Relativity*, in which the speed of light is constant for all observers *regardless of the motion of light source or observer*, is replaced with *Ritz's emission theory*, in which the speed of light is constant *with respect to the emitter.* As a consequence, in place of the 52 *elementary particles* and *antiparticles* in the Standard model of which only the electron and photon are stable, in New Physics the universe is composed of 14 *elementary particles. Elementary particles* are confined to those that have the possibility of being *observed*. There are no *quarks*, *gluons*, nor W^+, W^-, Z^0, or *Higgs bosons*. *Antiparticles* are simply the less stable *particles* of similar *mass* but of opposite *electric charge*. The notion of *antimatter* and problem of asymmetry of *matter* and *antimatter* in the visible universe are eliminated. See Underwood, T. G. (October, 2024). *New Physics*.]

Lubin, L. M. & Sandage, A. (June, 2001). *The Tolman Surface Brightness Test for the Reality of the Expansion*, was included because it was one of very few recent "mainstream" articles which claimed to evaluate *Zwicky's "tired-light" theory*, but in fact it did not. Although Allan Sandage was one of the most influential astronomers of the 20th century, he was 75 years old at this time. It was very disappointing. It omitted an expansion factor, then failed to include the *"tired-light" factor* representing the linear loss of energy as the *photons* interact with *electrons* and other matter as they travel through intergalactic space. The evidence for an *expanding universe* which it provided based on *Tolman surface brightness test* was demonstrated by Lerner, E. J., Falomo, R., & Scarpa, R. (2014). *UV surface brightness of galaxies from the local universe to z ~ 5*, to be in error, and, when calculated correctly, *concluded that far from disproving a non-expanding cosmology, data by Lubin and Sandage agreed very well with predictions for a static Euclidean universe*. However, this paper provides some useful insights on this analysis, and on the deeply ingrained biases at work.

Mamas, D. L. (2010). *An explanation for the cosmological redshift*, is a bizarre attempt, by an unrecognized author based in Florida (with a PhD in physics from UCLA), to reformulate *Zwicky's "tired-light" theory* in terms of a simplistic pseudo-classical notion of a *photon* as an electromagnetic wave which causes a *free electron* to oscillate and reradiate. It is included here together with Mamas, D. L. (May, 2015). *Cosmological redshift model now experimentally confirmed*, as it provides useful background information in the absence of any mainstream paper. It also provides calculations that suggest that the *cosmological redshift*, based on the quantum mechanical formulation of

Zwicky's "tired-light" theory and estimates of the density of *free electrons* in intergalactic space, support a *static* rather than an *expanding universe*, on which the *Big Bang* origin of the universe is based. It incorrectly assumes, as Zwicky stated, that *Compton scattering* suffers from the blurring of images.

"The effect of a strong magnetic field on a dielectric" describes how Mamas, D. L. (2010) is reminiscent of a recent explanation by the author of the *Faraday effect*, in which, when addressed in terms of *non-relativistic quantum electrodynamics* (or New Physics) "the *electrons* in the *dielectric* under the influence of the strong *magnetic field* line up according to their *spin*. The motion thus effected will be *circular*; and rotating *electric charges* create a *magnetic field*. So, the circularly moving *electrons* will create their own *magnetic field* in addition to the external *magnetic field* on them.

Shaoa, M-H. (April, 2013). *The energy loss of photons and cosmological redshift*, is another rather bizarre article in which Zwicky's *"tired-light"* theory is presented in terms of the simplistic interpretation of the wave-particle nature of the *photon* and classical electromagnetic theory. This is rather surprising as it was published by Xinjiang Astronomical Observatory, Key Laboratory of Radio Astronomy, People's Republic of China, though not by a mainstream theoretical physics institution. It is included for the same reasons as Mamas, D. L. (2010).

Lerner, E. J., Falomo, R., & Scarpa, R. (2014). *UV surface brightness of galaxies from the local universe to z ~ 5*, is an important article that demonstrates errors in previous claims that have been made that the Tolman test provides compelling evidence against a *static model* for the universe. This was reconsidered by adopting a *static Euclidean universe* (SEU) with *a linear Hubble relation at all z* (which is not the *standard Einstein–de Sitter model*), resulting in a relation between *flux* and *luminosity* that is virtually indistinguishable from the one used for ΛCDM models. Based on the analysis of the UV SB of luminous disk galaxies from HUDF and GALEX datasets, reaching from the local universe to z ~ 5, it was shown that the *surface brightness* (SB) *remains constant as expected in a static universe.*

In particular, it shows that the conclusions in Lubin & Sandage *are not supported by the data* for two main reasons: (1) for the *static scenario,* Lubin and Sandage set the distance to d = (c/H_0) ln(1 + z), *which is valid only for the Einstein-de Sitter static case.* This is not the cosmology being tested where the *Hubble relation* is hypothesized to be d = cz/H_0 at all *redshift.* The conversion factors to transform arc seconds to pc in the *non-expanding model* are therefore different; (2) *the local sample includes only first rank cluster galaxies, while the high-z sample includes about 20 normal galaxies in each of three different clusters.* This means that their distant galaxies are on average smaller and less luminous, and therefore are not directly comparable to local ones because of the well-known *absolute magnitude-SB relation. It concludes that far from disproving a non-expanding cosmology, data by Lubin and Sandage agree very well with predictions for a static Euclidean universe.*

The current volume concludes that *there was no Big Bang, the universe is not just 13.8 billion years old,* but *is indefinitely old, and in a steady state, not expanding.*

Hubble, E. (April, 1929). A relation between distance and radial velocity among extra-galactic nebulae.

Proc. Nat. Acad. Sci., 15, 3, 168-73; https://doi.org/10.1073/pnas.15.3.168.

Mount Wilson Observatory, Carnegie Institute of Washington

Communicated: January 17, 1929.

Hubble's famous paper in which he used the strong direct relationship between a classical Cepheid variable's *luminosity* and *pulsation period* for scaling galactic and extragalactic *distances* and confirmed that the *redshift* of a galaxy, expressed as its *radial velocity*, increases with its *distance* from Earth, a behavior that became known as Hubble's law.

———————————

Determinations of the motion of the sun with respect to the extra galactic nebulae have involved a K term of several hundred kilometers which appears to be variable. Explanations of this paradox have been sought in a correlation between apparent *radial velocities* and *distances*, but so far, the results have not been convincing. The present paper is a re-examination of the question, based on only those nebular distances which are believed to be fairly reliable.

Distances of extra-galactic nebulae depend ultimately upon the application of *absolute-luminosity* criteria to involved stars whose types can be recognized. These include, among others, Cepheid variables, novae, and blue stars involved in *emission nebulosity*. Numerical values depend upon the zero point of the *period-luminosity relation* among Cepheids, the other criteria merely check the order of the distances. This method is restricted to the few nebulae which are well resolved by existing instruments. A study of these nebulae, together with those in which any stars at all can be recognized, indicates the probability of an approximately uniform upper limit to the *absolute luminosity* of stars, in the late-type spirals and irregular nebulae at least, of the order of M (photographic) = – 6.3.[1]

[1] Mt. Wilson Contr., No. 324 (1926). *Astroph. J.*, Chicago, Ill., 64, 321.

The *apparent luminosities* of the brightest stars in such nebulae are thus criteria which, although rough and to be applied with caution, furnish reasonable estimates of the *distances* of all extra-galactic systems in which even a few stars can be detected.

Finally, the nebulae themselves appear to be of a definite order of *absolute luminosity*, exhibiting a range of four or five magnitudes about an average value M (visual) = – 15.2.[1] The application of this statistical average to individual cases can rarely be used to advantage, but where considerable numbers are involved, and especially in the various clusters of nebulae, *mean apparent luminosities* of the nebulae themselves offer reliable estimates of the *mean distances*.

Radial velocities of 46 extra-galactic nebulae are now available, but individual distances are estimated for only 24.

> [The *radial velocity* or line-of-sight velocity of a target with respect to an observer is the rate of change of the vector displacement between the two points. In astronomy, *radial velocity* is often measured to the first order of approximation by Doppler spectroscopy.]

For one other, N. G. C. 3521, an estimate could probably be made, but no photographs are available at Mount Wilson.

The data are given in table 1. The first seven *distances* are the most reliable, depending, except for M 32 the companion of M 31, upon extensive investigations of many stars involved. The next thirteen *distances*, depending upon the criterion of a uniform upper limit of *stellar luminosity*, are subject to considerable probable errors but are believed to be the most reasonable values at present available. The last four objects

23

appear to be in the Virgo Cluster. The *distance* assigned to the cluster, 2 x 10^6 parsecs, is derived from the distribution of *nebular luminosities*, together with *luminosities* of stars in some of the later-type spirals, and differs somewhat from the Harvard estimate of ten million light years[2].

[2] (1926). *Harvard Coll. Obs. Circ.*, 294.

The data in the table indicate a linear correlation between *distances* and *velocities*, whether the latter are used directly or corrected for solar motion, according to the older solutions. This suggests a new solution for the solar motion in which the *distances* are introduced as coefficients

Table 1.

Nebulae whose distances have been estimated from stars involved or from mean luminosities in a cluster.

OBJECT	m_s	r	v	m_t	M_t
S. Mag.	..	0.032	+ 170	1.5	−16.0
L. Mag.	..	0.034	+ 290	0.5	17.2
N. G. C. 6822	..	0.214	− 130	9.0	12.7
598	..	0.263	− 70	7.0	15.1
221	..	0.275	− 185	8.8	13.4
224	..	0.275	− 220	5.0	17.2
5457	17.0	0.45	+ 200	9.9	13.3
4736	17.3	0.5	+ 290	8.4	15.1
5194	17.3	0.5	+ 270	7.4	16.1
4449	17.8	0.63	+ 200	9.5	14.5
4214	18.3	0.8	+ 300	11.3	13.2
3031	18.5	0.9	− 30	8.3	16.4
3627	18.5	0.9	+ 650	9.1	15.7
4826	18.5	0.9	+ 150	9.0	15.7
5236	18.5	0.9	+ 500	10.4	14.4
1068	18.7	1.0	+ 920	9.1	15.9
5055	19.0	1.1	+ 450	9.6	15.6
7331	19.0	1.1	+ 500	10.4	14.8
4258	19.5	1.4	+ 500	8.7	17.0
4151	20.0	1.7	+ 960	12.0	14.2
4382	..	2.0	+ 500	10.0	16.5
4472	..	2.0	+ 850	8.8	17.7
4486	..	2.0	+ 800	9.7	16.8
4649	..	2.0	+1090	9.5	17.0
Mean					−15.5

m_s = photographic magnitude of brightest stars involved.
r = *distance* in units of 10^6 parsecs. The first two are Shapley's values.

v = measured *velocities* in km./sec. N. G. C. 6822, 221, 224 and 5457 are recent determinations by Humason.

m_t = Holetschek's visual magnitude as corrected by Hopmann. The first three objects were not measured by Holetschek, and the values of m_t represent estimates by the author based upon such data as are available.

M_t = total visual absolute magnitude computed from m_t and r.

of the K term, i. e., the *velocities* are assumed to vary, directly with the distances, and hence K represents the velocity at unit distance due to this effect. The equations of condition then take the form

$$rK + X\cos \alpha \cos \delta + Y \sin \alpha \cos \delta + Z \sin \delta = v.$$

Two solutions have been made, one using the 24 nebulae individually, the other combining them into 9 groups according to proximity in direction and in distance. The results are

	24 OBJECTS	9 GROUPS
X	$-\ 65 \pm 50$	$+\ \ \ 3 \pm\ \ 70$
Y	$+226 \pm 95$	$+230 \pm 120$
Z	-195 ± 40	$-133 \pm\ \ 70$
K	$+465 \pm 50$	$+513 \pm\ \ 60$ km./sec. per 10^6 parsecs.
A	$286\degree$	$269\degree$
D	$+\ 40\degree$	$+\ 33\degree$
V_0	306 km./sec.	247 km./sec.

For such scanty material, so poorly distributed, the results are fairly definite. Differences between the two solutions are due largely to the four Virgo nebulae, which, being the most distant objects and all sharing the peculiar motion of the cluster, unduly influence the value of K and hence of Vo. New data on more distant objects will be required to reduce the effect of such peculiar motion. Meanwhile round numbers, intermediate between the two solutions, will represent the probable order of the values. For instance, let A = 277°, D = +36° (Gal. long. = 32°, lat. = +18°), Vo = 280 km./sec., K = +500 km./sec. per million par secs. Mr. Stromberg has very kindly checked the general order of these values by independent solutions for different groupings of the data.

A constant term, introduced into the equations, was found to be small and negative. This seems to dispose of the necessity for the old constant K term. Solutions of this sort have been published by Lundmark[3], who replaced the old K by $k + lr + mr^2$.

[3] (1925). *Mon. Not. R. Astr. Soc.*, 85, 865-94.

His favored solution gave $k = 513$, as against the former value of the order of 700, and hence offered little advantage.

Table 2.
Nebulae whose distances are estimated from radial velocities.

OBJECT		v	v_s	r	m_t	M_t
N. G. C.	278	+ 650	−110	1.52	12.0	−13.9
	404	− 25	− 65	..	11.1	..
	584	+1800	+ 75	3.45	10.9	16.8
	936	+1300	+115	2.37	11.1	15.7
	1023	+ 300	− 10	0.62	10.2	13.8
	1700	+ 800	+220	1.16	12.5	12.8
	2681	+ 700	− 10	1.42	10.7	15.0
	2683	+ 400	+ 65	0.67	9.9	14.3
	2841	+ 600	− 20	1.24	9.4	16.1
	3034	+ 290	−105	0.79	9.0	15.5
	3115	+ 600	+105	1.00	9.5	15.5
	3368	+ 940	+ 70	1.74	10.0	16.2
	3379	+ 810	+ 65	1.49	9.4	16.4
	3489	+ 600	+ 50	1.10	11.2	14.0
	3521	+ 730	+ 95	1.27	10.1	15.4
	3623	+ 800	+ 35	1.53	9.9	16.0
	4111	+ 800	− 95	1.79	10.1	16.1
	4526	+ 580	− 20	1.20	11.1	14.3
	4565	+1100	− 75	2.35	11.0	15.9
	4594	+1140	+ 25	2.23	9.1	17.6
	5005	+ 900	−130	2.06	11.1	15.5
	5866	+ 650	−215	1.73	11.7	−14.5
Mean					10.5	−15.3

The residuals for the two solutions given above and below average 150 and 110 km./sec. and should represent the average peculiar motions of the individual nebulae and of the groups, respectively. In order to exhibit the results in a graphical form, the solar motion has been eliminated from the observed *velocities* and the remainders, the *distance* terms plus the residuals, have been plotted against the *distances*. The run of the residuals is about as smooth as can be expected, and in general the form of the solutions appears to be adequate.

The 22 nebulae for which *distances* are not available can be treated in two ways. First, the *mean distance* of the group derived from the mean apparent magnitudes can be compared with the *mean of the velocities* corrected for solar motion. The result, 745 km./sec. for a distance of 1.4 x 10⁶ parsecs, falls between the two previous solutions and indicates a value for K of 530 as against the proposed value, 500 km./sec.

Secondly, the scatter of the individual nebulae can be examined by assuming the relation between *distances* and *velocities* as previously determined. *Distances* can then be calculated from the *velocities* corrected for solar motion, and absolute magnitudes can be derived from the apparent magnitudes. The results are given in table 2 and may be compared with the distribution of absolute magnitudes among the nebulae in table 1, whose *distances* are derived from other criteria.
N. G. C. 404 can be excluded, since the observed velocity is so small that the peculiar motion must be large in comparison with the *distance* effect. The object is not necessarily an exception, however, since a

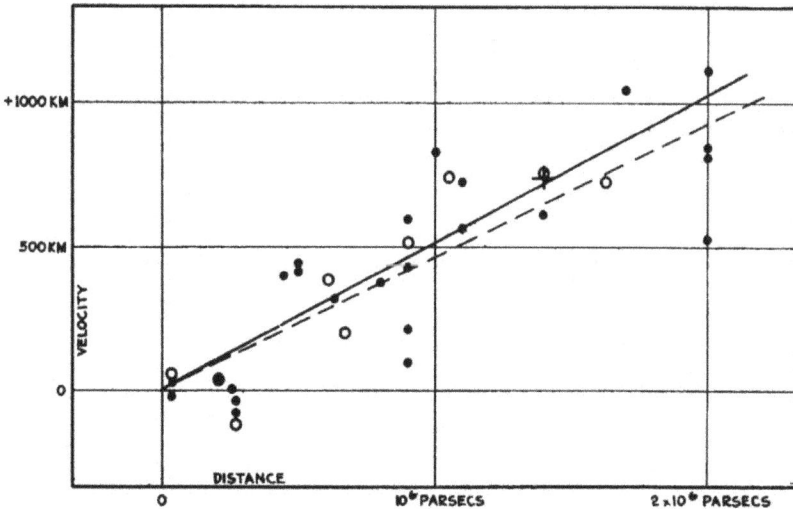

FIGURE 1. Velocity-Distance Relation among Extra-Galactic Nebulae. Radial velocities, corrected for solar motion, are plotted against distances estimated from involved stars and mean luminosities of nebulae in a cluster. The black discs and full line represent the solution for solar motion using the nebulae individually; the circles and broken line represent the solution combining the nebulae into groups; the cross represents the mean velocity corresponding to the mean distance of 22 nebulae whose distances could not be estimated individually.

distance can be assigned for which the peculiar motion and the absolute magnitude are both within the range previously determined. The two mean magnitudes, – 15.3 and – 15.5, the ranges, 4.9 and 5.0 mag., and the frequency distributions are closely similar for these two entirely independent sets of data; and even the slight difference in mean magnitudes can be attributed to the selected, very bright, nebulae in the Virgo Cluster. This entirely unforced agreement supports the validity of the *velocity-distance* relation in a very evident matter. Finally, it is worth recording that the frequency distribution of absolute magnitudes in the two tables combined is comparable with those found in the various clusters of nebulae.

The results establish a roughly linear relation between *velocities* and *distances* among nebulae for which velocities have been previously published, and the relation appears to dominate the distribution of *velocities*. In order to investigate the matter on a much larger scale, Mr. Humason at Mount Wilson has initiated a program of determining *velocities* of the most distant nebulae that can be observed with confidence. These, naturally, are the brightest nebulae in clusters of nebulae. The first definite result[4], v = + 3,779 km./sec. for N. G. C. 7619, is thoroughly consistent with the present conclusions.

[4] (1929). *These proceedings*, 15, 167.

Corrected for the solar motion, this *velocity* is +3,910, which, with K = 500, corresponds to a *distance* of 7.8×10^6 parsecs. Since the apparent magnitude is 11.8, the absolute magnitude at such a *distance* is – 17.65, which is of the right order for the brightest nebulae in a cluster. A preliminary *distance*, derived independently from the cluster of which this nebula appears to be a member, is of the order of 7×10^6 parsecs.

New data to be expected in the near future may modify the significance of the present investigation or, if confirmatory, will lead to a solution having many times the weight. For this reason, it is thought premature to discuss in detail the obvious consequences of the present results. For example, if the solar motion with respect to the clusters represents the *rotation* of the galactic system, this motion could be subtracted from the results for the nebulae and the remainder would represent the motion of the galactic system with respect to the extra-galactic nebulae.

The outstanding feature, however, is the possibility that the *velocity-distance* relation may represent the *de Sitter effect*, and hence that numerical data may be introduced into discussions of the general curvature of space. *In the de Sitter cosmology, displacements of the spectra arise from two sources, an apparent slowing down of atomic vibrations and a general tendency of material particles to scatter.* The latter involves an acceleration and hence introduces the element of time. The relative importance of these two effects should determine the form of the relation between *distances* and observed *velocities*; and in this connection it may be emphasized that the linear relation found in the present discussion is a first approximation representing a restricted range in *distance*.

Edwin Powell Hubble (November 20, 1889 – September 28, 1953).

Hubble was an American astronomer. He played a crucial role in establishing the fields of extragalactic astronomy and observational cosmology. Hubble proved that many objects previously thought to be clouds of dust and gas and classified as "nebulae" were actually galaxies beyond the Milky Way. He used the strong direct relationship between a classical Cepheid variable's luminosity and pulsation period for scaling galactic and extragalactic distances. Hubble confirmed in 1929 that the recessional velocity of a galaxy increases with its distance from Earth, a behavior that became known as Hubble's law, although it had been proposed two years earlier by Georges Lemaître.

Hubble's name is most widely recognized for the Hubble Space Telescope, which was named in his honor, with a model prominently displayed in his hometown of Marshfield, Missouri.

Hubble was born in 1889 to Virginia Lee Hubble (née James) (1864–1934) and John Powell Hubble, an insurance executive, in Marshfield, Missouri, and moved to Wheaton, Illinois, in 1900. In his younger days, he was noted more for his athletic prowess than his intellectual abilities, although he did earn good grades in every subject except spelling. Edwin was a gifted athlete, playing baseball, football, and running track in both high school and college. He won seven first places and a third place in a single high school track and field meet in 1906, and he played a variety of positions on the basketball court from center to shooting guard. Hubble led the University of Chicago's basketball team to their first Big Ten Conference title in 1907.

Hubble's studies at the University of Chicago were concentrated on mathematics, astronomy and philosophy, which resulted in a bachelor of science degree by 1910. For a year he was also a student laboratory assistant for the physicist Robert Millikan, a future Nobel Prize winner. Hubble was a Rhodes Scholar, he spent three years at The Queen's

College, Oxford studying jurisprudence instead of science (as a promise to his dying father), and later added studies in literature and Spanish, eventually earning his master's degree.

In 1909, Hubble's father moved his family from Chicago to Shelbyville, Kentucky, so that the family could live in a small town, ultimately settling in nearby Louisville. His father died in the winter of 1913, while Edwin was still in England. In the following summer, Edwin returned home to care for his mother, two sisters, and younger brother, along with his brother William. The family moved once more to Everett Avenue, in Louisville's Highlands neighborhood, to accommodate Edwin and William.

Hubble was a dutiful son, who despite his intense interest in astronomy since boyhood, acquiesced to his father's request to study law, first at the University of Chicago and later at Oxford. In this time, he also took some math and science courses. After the death of his father in 1913, Edwin returned to the Midwest from Oxford but did not have the motivation to practice law. Instead, he proceeded to teach Spanish, physics and mathematics at New Albany High School in New Albany, Indiana, where he also coached the boys' basketball team. After a year of high-school teaching, he entered graduate school with the help of his former professor from the University of Chicago to study astronomy at the university's Yerkes Observatory, where he received his Ph.D. in 1921. His dissertation was titled "Photographic Investigations of Faint Nebulae". At Yerkes, he had access to its 40-inch refractor built in 1897, as well as an innovative 26-inch (61 cm) reflector

After the United States declared war on Germany in 1917 during World War I, Hubble rushed to complete his Ph.D. dissertation so he could join the military. Hubble volunteered for the United States Army and was assigned to the newly created 86th Division, where he served in the 2nd Battalion, 343rd Infantry Regiment. He rose to the rank of major, and was found fit for overseas duty on July 9, 1918; the 86th Division moved overseas, but never saw combat as it was broken up and its personnel used as replacements in other units. After the end of World War I, Hubble spent a year at Cambridge University, where he renewed his studies of astronomy.

In 1919, Hubble was offered a staff position at the Carnegie Institution for Science's Mount Wilson Observatory, near Pasadena, California, by George Ellery Hale, the founder and director of the observatory. Hubble remained on staff at Mount Wilson until his death in 1953. Shortly before his death, Hubble became the first astronomer to use the newly completed giant 200-inch (5.1 m) reflector Hale Telescope at the Palomar Observatory near San Diego, California.

Edwin Hubble's arrival at Mount Wilson Observatory, California, in 1919 coincided roughly with the completion of the 100-inch (2.5 m) Hooker Telescope, then the world's largest. At that time, the prevailing view of the cosmos was that the universe consisted entirely of the Milky Way Galaxy.

Hubble married Grace Lillian (Burke) Leib (1889–1980), daughter of John Patrick and Luella (Kepford) Burke, on February 26, 1924.

Using the Hooker Telescope at Mount Wilson, Hubble identified Cepheid variables, a standard candle discovered by Henrietta Leavitt. Comparing their apparent luminosity to their intrinsic luminosity gives their distance from Earth. Hubble found Cepheids in several nebulae, including the Andromeda Nebula and Triangulum Nebula. His observations, made in 1924, proved conclusively that these nebulae were much too distant to be part of the Milky Way and were, in fact, entire galaxies outside our own; thus today they are no longer considered nebulae.

Hubble's hypothesis was opposed by many in the astronomy establishment of the time, in particular by Harvard University–based Harlow Shapley. Despite the opposition, Hubble, then a thirty-five-year-old scientist, had his findings first published in The New York Times on November 23, 1924, then presented them to other astronomers at the January 1, 1925, meeting of the American Astronomical Society.

Hubble's results for Andromeda were not formally published in a peer-reviewed scientific journal until 1929.

Hubble's findings fundamentally changed the scientific view of the universe. Supporters state that Hubble's discovery of nebulae outside of our galaxy helped pave the way for future astronomers. Although some of his more renowned colleagues simply scoffed at his results, Hubble ended up publishing his findings on nebulae.

Hubble also devised the most commonly used system for classifying galaxies, grouping them according to their appearance in photographic images. He arranged the different groups of galaxies in what became known as the *Hubble sequence*.

Hubble went on to estimate the distances to 24 *extra-galactic nebulae*, using a variety of methods. In 1929 Hubble examined the relationship between these *distances* and their *radial velocities* as determined from their *redshifts*. All of his estimated distances are now known to be too small, by up to a factor of about 7. This was due to factors such as the fact that there are two kinds of Cepheid variables or confusing bright gas clouds with bright stars. However, his *distances* were more or less proportional to the true distances, and combining his distances with measurements of the *redshifts* of the galaxies by Vesto Slipher, and by his assistant Milton L. Humason, he found a roughly linear relationship between the *distances* of the galaxies and their *radial velocities* (corrected for solar motion), a discovery that later became known as *Hubble's law*.

This meant that the greater the distance between any two galaxies, the greater their relative speed of separation. If interpreted that way, Hubble's measurements on 46 galaxies lead to a value for the Hubble constant of 500 km/s/Mpc, which is much higher than the currently accepted values of 74 km/s/Mpc (cosmic distance ladder method) or 68 km/s/Mpc (CMB method) due to errors in their distance calibrations.

There were methodological problems with Hubble's survey technique that showed a deviation from flatness at large *redshifts*. In particular, the technique did not account for changes in luminosity of galaxies due to galaxy evolution. Earlier, in 1917, Albert Einstein had found that his newly developed *theory of general relativity* indicated that the universe must be either expanding or contracting. Unable to believe what his own equations were telling him, Einstein introduced a *cosmological constant* (a "fudge factor") to the equations to avoid this "problem". When Einstein learned of Hubble's *redshifts*, he immediately realized that the expansion predicted by *general relativity* must be real, and in later life, he said that changing his equations was "the biggest blunder of [his] life." In fact, Einstein apparently once visited Hubble and tried to convince him that the universe was expanding.

Yet the reason for the *redshift* remained unclear. Georges Lemaître had predicted on theoretical grounds based on *Einstein's equations for general relativity* the *redshift-distance relation*, and published observational support for it, two years before the discovery of Hubble's law. Although Hubble used the term "*velocities*" in his paper (and "*apparent radial velocities*" in the introduction), he later expressed doubt about interpreting these as real velocities. In 1931 he wrote a letter to the Dutch cosmologist Willem de Sitter expressing his opinion on the theoretical interpretation of the *redshift-distance* relation:

> "Mr. Humason and I are both deeply sensible of your gracious appreciation of the papers on velocities and distances of nebulae. We use the term *'apparent' velocities* to emphasize the empirical features of the correlation. The interpretation, we feel, should be left to you and the very few others who are competent to discuss the matter with authority."

In the 1930s, Hubble was involved in determining the distribution of galaxies and spatial curvature. These data seemed to indicate that the universe was flat and homogeneous, but there was a deviation from flatness at large redshifts. According to Allan Sandage,

"Hubble believed that his count data gave a more reasonable result concerning spatial curvature if the redshift correction was made assuming no recession. To the very end of his writings, he maintained this position, *favoring (or at the very least keeping open) the model where no true expansion exists*, and therefore that the redshift "represents a hitherto unrecognized principle of nature."

Hubble also discovered the asteroid 1373 Cincinnati on August 30, 1935. In 1936 he wrote The Observational Approach to Cosmology and The Realm of the Nebulae which explained his approaches to extra-galactic astronomy and his view of the subject's history.

Hubble also worked as a civilian for U.S. Army at Aberdeen Proving Ground in Maryland during World War II as the Chief of the External Ballistics Branch of the Ballistics Research Laboratory during which he directed a large volume of research in exterior ballistics which increased the effective firepower of bombs and projectiles. His work was facilitated by his personal development of several items of equipment for the instrumentation used in exterior ballistics, the most outstanding development being the high-speed clock camera, which made possible the study of the characteristics of bombs and low-velocity projectiles in flight. The results of his studies were credited with greatly improving design, performance, and military effectiveness of bombs and rockets. For his work there, he received the Legion of Merit award.

In December 1941, Hubble reported to the American Association for the Advancement of Science that results from a six-year survey with the Mt. Wilson telescope *did not support the expanding universe theory*. According to a Los Angeles Times article reporting on Hubble's remarks, "The nebulae could not be uniformly distributed, as the telescope shows they are, and still fit the explosion idea. Explanations which try to get around what the great telescope sees, he said, fail to stand up. The explosion, for example, would have had to start long after the earth was created, and possibly even after the first life appeared here." (Hubble's

35

estimate of what we now call the Hubble constant would put the Big Bang only 2 billion years ago.)

Hubble had a heart attack in July 1949 while on vacation in Colorado. He was cared for by his wife and continued on a modified diet and work schedule. He died of cerebral thrombosis (a blood clot in his brain) on September 28, 1953, in San Marino, California. No funeral was held for him, and his wife never revealed his burial site.

Hubble's papers comprising the bulk of his correspondence, photographs, notebooks, observing logbooks, and other materials, are held by the Huntington Library in San Marino, California. They were donated by his wife Grace Burke Hubble upon her death in 1980.

During Hubble's life the Nobel Prize in Physics did not cover work done in astronomy. Hubble spent much of the later part of his career attempting to have astronomy considered part of physics, instead of being a separate science. He did this largely so that astronomers—including himself—could be recognized by the Nobel Prize Committee for their valuable contributions to astrophysics. This campaign was unsuccessful in Hubble's lifetime, but shortly after his death, the Nobel Prize Committee decided that astronomical work would be eligible for the physics prize. However, the Nobel prize is not awarded posthumously.

Zwicky, F. (1929). On the Red Shift of Spectral Lines through Interstellar Space.

Proc. Natl. Acad. Sci., 15, 773-9; https://www.pnas.org/doi/pdf/10.1073/pnas.15.10.773.

Norman Bridge Laboratory of Physics, California Institute of Technology.

Communicated August 26, 1929.

Zwicky's famous paper in which he introduced his *"tired-light" theory*. When Edwin Hubble discovered a somewhat linear relationship between the *distance* to a galaxy and its *redshift* expressed as a velocity, Zwicky immediately pointed out that the correlation between the calculated distances of galaxies and their redshifts had a discrepancy too large to fit in the distance's error margins. *He proposed that the reddening effect was not due to motions of the galaxy, but to an unknown phenomenon that caused photons to lose energy as they traveled through space.* He considered the most likely candidate process to be a drag effect in which photons transfer momentum to surrounding masses through gravitational interactions. He rejected explanations involving the expansion of space, and incorrectly stated that *Compton scattering* suffers from the blurring of images.

A. *Introduction*. - It is known that very distant *nebulae*, probably galactic systems like our own, show remarkably high receding velocities whose magnitude increases with the distance. This curious phenomenon promises to provide some important clues for the future development of our cosmological views. It may be of advantage, therefore, to point out some of the principal facts which any cosmological theory will have to account for. Then a brief discussion will be given of different theoretical suggestions related to the above effect. Finally, *a new effect of masses upon light will be suggested* which is a sort of gravitational analogue of the *Compton effect*.

B. *Discussion of the Observational Facts.* - (1) E. Hubble[1] has shown recently that *the correlation between the apparent velocity of recession and the distance is roughly linear*, corresponding to 500 km./sec. per 10^6 parsecs.

[1] Hubble, E. (April, 1929). A relation between distance and radial velocity among extra-galactic nebulae. *Proc. Nat. Acad. Sci.*, 15, 3, 168-73; https://doi.org/10.1073/pnas.15.3.168.

Large deviations occur for the nearest nebulae, which may be attributed to their peculiar motions. The most recent observations by M. Humason[2] seem to indicate that for very large distances (50×10^6 light years) the individual deviations become so great (3,000 km./sec. out of 8,000 km./sec.) that they hardly can be due to peculiar motions and must, therefore, be accounted for in some other way.

[2] I am indebted to Mr. M. Humason and to Dr. E. Hubble for private information.

(2) *The relative shift of frequency $\Delta v/v$ representing the velocity of recession is apparently independent of the frequency.* The available range in the spectrum is not very large, however. Some exceptions have been found, suggesting that $\Delta v/v$ for H_β, is somewhat greater than for H_γ.

(3) No appreciable *absorption* or *scattering* of light can be related to the above shift of spectral lines.

(4) The optical image of an extragalactic nebula seems to be as well defined as can be expected from the resolving power of the telescopes. The distance apparently is only geometrically involved and *no additional blurring of the images occurs* due to some such process as multiple scattering and superposition of incoherent light beams.

(5) The *spectral (absorption) lines* obtained from these nebulae are not very well defined, but no systematic investigation of their shape has been carried out. According to recent observations by M. Humason the width of the lines ranges between 4 A and 7 A for M_{32} and M_{31} the two Andromeda nebulae.

(6) Extrapolating from Hubble's relation to objects in our own galactic system, the velocity of recession would become so small (5 km./sec. for 10,000 parsecs) that it would escape observation. *The theoretical considerations proposed by the author in the following made it probable that an appreciable effect should also be observed in our galaxy.* This suggestion was tested by Dr. ten Bruggencate, whose work will be published shortly. His essential result is that *the velocity of recession of the globular clusters is a function of the galactic latitude, increasing with decreasing latitude*[4].

[4] A paper by E. von der Pahlen and E. Freundlich in the *Publikationen des Astrophysikalischen Observatoriums zu Potsdam*, 86, Bd. 26, Heft 3, should be mentioned in this connection. Fig. 6 on page 44 of this paper also shows a correlation between the radial velocity of globular clusters and the galactic latitude. The authors, however, interpret it as being due to real motions.

We proceed now in discussing different theoretical possibilities of accounting for the phenomenon described above.

C. *de Sitter's Universe*. - It has been pointed out by de Sitter that the special type of a space proposed by himself as representing our universe would imply on the average a velocity of recession of the far distant nebulae. But the linear relation of Hubble's can only be obtained by making some additional assumptions about the distribution of the nebulae. For more detailed information, we refer to a recent paper by R. C. Tolman[3].

[3] Tolman, R. C. (1929). On the astronomical implications of the de Sitter line element for the universe. *Astrophys. J.*, 49, 245; https://articles.adsabs.harvard.edu/cgi-bin/nph-iarticle_query?1929ApJ....69..245T&defaultprint=YES&page_ind=0 &filetype=.pdf: "Abstract: Part I. Nature of *De Sitter universe*.—Using the form of line element proposed by Eddington, exact expressions are obtained/or the motion of particles and light-rays in the *De Sitter universe*, which include the shape of the orbit, the velocity of motion in the orbit, the acceleration in the orbit, and, for purely radial motion,

the integrated equations of motion. Part II. Relation to astronomical measurements.—The relation is discussed between the co-ordinate r and the determinations of radial distance that would result from direct measurement with meter sticks, from measurements of parallax, from those of the apparent luminosity of distant objects, and from those of the apparent extension of distant objects. A relation between the proper density of distribution of objects and coordinate density is obtained. Exact expressions are derived for the *Doppler effect* in terms of *distances* and *velocities* or in terms of distances and orbital parameters, both for the case that the observer is at the origin and for the case that the source is at the origin of co-ordinates. The *Doppler effect* at perihelion, and the range of distance and range of time during which an approaching source would produce the reversed positive *Doppler effect*, are discussed. Part III. Relation to astronomical findings.—The known facts as to the distances and *Doppler effects* for the extra-galactic nebulae are discussed on the assumption that these objects can be regarded as free particles in a *De* Sitter universe. Four possible hypotheses that would reconcile the presence of nebulae within the range of observation with the scattering tendency which exists in the *De Sitter universe* are presented and discussed. The most natural of these hypotheses is that of continuous entry in accordance with which nebulae are continually entering as well as leaving the range of observation. This hypothesis leads to a *Doppler effect* which tends to be positive and to increase with *distance*, but an overwhelming predominance of positive *Doppler effects* and a linear increase with *distance* could only be secured by placing additional conditions on the parameters for the orbits which do not appear inevitable. The other hypotheses appear less natural. They can be made to agree with the experimental facts by the introduction of ad hoc assumptions. The conclusion is drawn that the *De Sitter* line element does not afford a simple and unmistakably evident explanation of our present knowledge of the distribution, distances, and *Doppler effects* for the extra-galactic nebulae".

Admitting that *de Sitter*'s explanation accounts for the facts listed above in the sections B_1 to B_6, *a correlation of the type B_6 for our own galaxy would present an almost unsurmountable obstacle for any theory based on geometry only.*

D. *The Compton-Doppler Effect on Free Electrons.* - We know from different sources, that *there exist very dilute gaseous masses distributed all over the interstellar spaces.* The observations of the steady Ca^+ and Na absorption lines provide one of the most direct proofs of this fact. These observations also show that some of the atoms occur as ions. It may be concluded, therefore, that an adequate number of *free electrons* be present. One then might expect that the light coming from distant nebulae would undergo a shift to the red by *Compton effect* on those *free electrons.*

[*Compton scattering* (or the *Compton effect*) is the *quantum theory* of *high frequency photons scattering* following an interaction with a *charged particle*, usually an *electron.* Specifically, when the *photon* hits *electrons*, it releases loosely bound *electrons* from the outer valence shells of atoms or molecules.

The effect was discovered in 1923 by Arthur Holly Compton while researching the scattering of X-rays by light elements, and earned him the Nobel Prize for Physics in 1927. The *Compton effect* significantly deviated from dominating classical theories, using both special relativity and quantum mechanics to explain the interaction between high frequency photons and charged particles.

Photons can interact with *matter* at the *atomic level* (e.g. *photoelectric effect* and *Rayleigh scattering*), at the *nucleus*, or with just an *electron. Pair production* and the *Compton effect* occur at the level of the *electron.* When a *high frequency photon* scatters due to an interaction with a *charged particle, there is a decrease in the energy of the photon* and thus, an increase in its *wavelength.* This tradeoff between *wavelength* and *energy* in response to the collision is the *Compton effect.* Because of conservation of energy, the lost energy from the *photon* is transferred to the recoiling particle.

Rayleigh scattering is the scattering or deflection of light, or other electromagnetic radiation, by particles with a size much smaller than the wavelength of the radiation. For light frequencies well below the resonance frequency of the scattering medium (normal dispersion regime), *the amount of scattering is inversely proportional to the fourth power of the wavelength* (e.g., a blue color is scattered much more than a red color as light propagates through air). The phenomenon is named after the 19th-century British physicist Lord Rayleigh (John William Strutt).

Rayleigh scattering results from the *electric polarizability* of the particles. The oscillating *electric field* of a light wave acts on the charges within a particle, causing them to move at the same frequency. *The particle, therefore, becomes a small radiating dipole whose radiation we see as scattered light.* The particles may be individual *atoms* or *molecules*; it can occur when light travels through transparent solids and liquids, but is most prominently seen in gases]

Now the admissible deflection in one single process is very small, the angular size of the nebulae being indeed less than one degree of arc. For the change in *wave-length* $\Delta\lambda$ by a single *Compton scattering* within the above angle, one obtains then $\Delta\lambda \leq 3 \times 10^{-13}$ cm., so that a great number of collisions between the light quanta and the electrons are necessary in order to produce a change $\Delta\lambda \sim 1$ to 100 A. *But then the light scattered in all directions would make the interstellar space intolerably opaque which disposes of the above explanation.*

[Zwicky's dismissal of *Compton scattering* appears to be premature. See Compton (May, 1923) below. In *Compton scattering*, the incoming *gamma-ray photon* is *deflected* through an angle θ with respect to its original direction. The *photon* transfers a portion of its *energy* to the *electron* (assumed to be initially at rest), which is then known as a *recoil electron*, or a *Compton electron*. The *energy* transferred to the *electron* can vary from zero to a large fraction of the *gamma-ray energy*. The *Compton scattering* probability decreases as the *photon energy* increases and is directly proportional to the number of *electrons* per gram, which only varies by 20% from the lightest to the heaviest elements (except for hydrogen). See "Interactions of Photons with Matter" below.]

It is possible, of course, that a great number of the *electrons* possess very high speed. This is suggested by the existence of the cosmic radiation. In this case, an appreciable shift of frequency may be produced by one collision. But *still the difficulty of obtaining too much scattered light in all directions can hardly be avoided*, at least if use is made of our present knowledge of the intensity distribution due to *Compton effect*. Also, it is evident that *any explanation based on a scattering process like the Compton effect or the Raman effect, etc., will be in a hopeless position regarding the good definition of the images as mentioned under B₄.*

[The *Raman effect* or *Raman scattering* is the inelastic scattering of *photons* by *matter*, meaning that there is both an *exchange of energy* and a *change in the light's direction*. Typically, this effect involves vibrational energy being gained by a molecule as incident *photons* from a visible laser are shifted to lower *energy*.]

E. *The Usual Gravitational Shift of Spectral Lines*. - One might expect a shift of spectral lines due to the difference of the static gravitational potential at different distances from the center of a galaxy. *This effect, of course, has no relation to the distance of the observed galaxy from our own system and, therefore, cannot provide any explanation of the phenomenon discussed in this paper.* But it might have some bearing on the width of the observed spectral lines as light coming from different points of the distant galaxy will show varying-shifts. To get an estimate, we assume the nebulae to be a sphere of the radius $R = 2 \times 10^4$ light years and of a uniform density 10^{-20} gr./cm.$^3 > \rho > 10^{-24}$ gr./cm.^{3}. Then we have for the *gravitational potential* $\Phi(r)$ this relation:

$\Delta\Phi = \Phi(R) - \Phi(0) = 2\pi \, fpR^2$ where $f = 6.68 \times 10^{-8}$ is the *universal gravitational constant*. And for the above limits of ρ, $8 \times 10^{-8} < \Delta v/v = \varphi/c^2 < 8 \times 10^{-4}$ (240 km/sec.) where c is the velocity of light. This effect also might cause a violet shift of the light traveling from the outer regions of our galaxy toward the center.

F. *The Gravitational "Drag" of Light.* - According to the *relativity theory*, a light quantum $\hbar v$ has an *inertial* and a *gravitational mass* $\hbar v/c^2$. It should be expected, therefore, that a quantum $\hbar v$ passing a *mass* M will not only be deflected but it will also transfer *momentum* and *energy* to the mass M and make it recoil. During this process, the light quantum will change its *energy* and, therefore, its *frequency*. It is hardly possible to give a completely satisfactory theory of this gravitational analogue of the *Compton effect*, without making use of the *general theory of relativity*. But a rough idea of the nature and the magnitude of the effect may be obtained in the following way.

Suppose a mass *m* to be traveling on a straight line (x-axis) with a uniform velocity υ. The mass M is located at the point P(x,y). If υ is sufficiently large, then the actual path will differ only little from the straight line. The force F acting between *m* and M can then be obtained in the first approximation by assuming that *m* is traveling along the x-axis with the constant initial velocity υ. If *m* for t = 0 at x = 0, then

$$F_x = -\partial\Phi/\partial x \, M \qquad \Phi(x,y,t) = -fm/\sqrt{\{y^2 + (x - \upsilon t)^2\}}$$
$$(\Phi = \text{gravitational potential at P.})$$

The x component of the momentum ΔG transferred to M during the time T is

$$\Delta G_x = +\int_0^T F_x \, dt = -M \int_0^T \partial\Phi/\partial x \, dt = M/\upsilon \int_0^T \partial\Phi/\partial t \, dt$$
$$= M/\upsilon \, [\Phi(T) - \Phi(0)].$$

If from t = 0 to t = T the particle has traveled the distance L = υT, then

$$\Delta G_x = -fmM/\upsilon \, [1/\sqrt{\{(L-x)^2 + y^2\}} - 1/\sqrt{(x^2 + y^2)}]$$
$$= fmM/\upsilon \, g_0(x,y).$$

Suppose that matter is distributed all over space with a uniform *density* ρ and put $M = \rho. 2\pi y \, dy dx$, then the x-component of the total *momentum* lost by m will be

$$G_x = \lim_{D=\infty} fm/\upsilon \int_{-D}^{D+L} \int_0^D 2\pi y\rho \ g_0(x,y) \ dx \ dy = 0.$$

An exchange of *momentum* in the x-direction results in this case only if we consider the actually occurring deflection of m from the straight path. This is a second order effect.

In the above calculation, we have assumed that the *gravitational interaction* is transmitted instantaneously. Let us consider now the case when gravity waves travel with the velocity of light c. Then we have according to the *theory of the retarded potentials*,

> [*Retarded potentials* are the *electromagnetic potentials* for the *electromagnetic field* generated by time-varying *electric current* or *charge distributions* in the past. The fields propagate at the speed of light c, so the delay of the fields connecting cause and effect at earlier and later times is an important factor: the signal takes a finite time to propagate from a point in the *charge* or *current* distribution (the point of cause) to another point in space (where the effect is measured).]

$$\Phi(x,y,t) = - fm/r(1 - \upsilon_r/c) \, | \, _{t'=t-r/c}$$

where the expression on the right has to be taken at the time of *emission* t' of an *action* reaching P at the time t. For the distance r of m and M at t' we have,

$$r^2 = y^2 + [x - \upsilon(t - r/c)]^2$$

and

$$\upsilon_r = [x - \upsilon(t - r/c)]\upsilon/r.$$

We notice again

$$\partial\Phi/\partial x = - 1/\upsilon \ \partial\Phi/\partial t$$

and

$$\Delta G_x = - M \int_{t_0}^{T} \partial \Phi / \partial x \; dt = M/\upsilon \int_{t_0}^{T} \partial \Phi / \partial t \; dt$$
$$= M/\upsilon \; [\Phi(t_T) - \Phi(t_0)].$$

The disturbance caused by the motion of m from $x = 0$, $t = 0$ to $t = T$, $x = L = \upsilon T$ is acting on M from $t_0 = \sqrt{(x^2 + y^2/c)}$ to $t_T = T + 1/c \; \sqrt{\{(L - x)^2 + y^2\}}$. We obtain therefore,

$$\Delta G_x = - fmM/\upsilon \; [1/\sqrt{\{(L - x)^2 + y^2 + (L - x)\upsilon/c\}} - 1/\sqrt{(x^2 + y^2 - x\upsilon/c)}]$$

Developing for $\upsilon/c \ll 1$

$$\Delta G_x = fmM/\upsilon \; [g_0(x,y) + g_1(x,y) \; v/c + \dots]$$
$g_0(x,y)$ is the same function as above
$$g_1(x,y) = (L - x)/\{(L - x)^2 + y^2\} + x/(x^2 + y^2).$$

Assuming again a uniform distribution of matter in space, we put $M = 2\pi y \; dydx.\rho$. The integration over g_0 gives zero as before and (for $D \gg L$)

$$2\pi \int_{-D}^{D+L} \int_{0}^{D} \; g_1(x,y)y \; dxdy = 2\pi lg2LD.$$

ΔG_x corresponding to the matter in a region $-D < x < D + L$ and $0 < y < D$ is therefore

$$\Delta G_x = 1.4\pi \; fm\rho LD/c.$$

In regard to D, it must be remarked that it should be as large as the dimension of the space over which masses are distributed, if those masses are regarded as independent from each other. But the masses are in reality coupled by gravitational forces and the effect of an external perturbation upon them must be computed by considering the system of the far distant masses as a whole. *The correct theory will probably have to be worked out in terms of absorption of gravitational waves.* But I think it may be safely assumed that the distance D in which the perturbing effect of the moving mass m begins to fade out *is very large* compared with the mutual distances of the single masses M in which matter is essentially concentrated.

Going over to the case of *light*, we have $\upsilon = c$ and $m = \hbar v/c^2$. We conclude by analogy that a relation of the above type still is valid, especially as it can be derived by simply using dimensional reasoning. Light traveling a distance L then would lose the *momentum*

$$\Delta(\hbar v/c) = 1.4\pi f\rho DL/c \times \hbar v/c^2 \text{ and } \Delta v/v = 1.4\pi f\rho DL/c^2.$$

Let us compare this result with the observations.

For the total space investigated, the possible limits for ρ are according to E. Hubble 10^{-26} gr./cm.$^3 > \rho > 10^{-31}$. The mutual distance l of the *galactic systems* being of the order $l = 10^6$ parsecs, we may assume D for instance of the order $1000\ l = 3 \times 10^{27}$ cm. Then $\Delta v/v$ for $L = 10^6$ parsecs according to our formula will be in the limits $3 \times 10^{-2} > \Delta v/v > 3 \times 10^{-7}$. From Hubble's linear relation, we have $\Delta v/v \sim 1/600$ for the same L. *In view of this agreement in order of magnitude, a further elaboration of the theory seems to be worthwhile.*

Applying the above theory to globular clusters of our own galaxy *it would be essential to take into account the actual mass distribution.* We will, however, obtain an estimate of the order of magnitude by taking L = 15,000 parsecs, D = 1000 l with l = 1 parsec for the mutual average distance of the stars. The limits for the density are 10^{-20} gr./cm.$^3 > \rho_g > 10^{-24}$ gr./cm.^{3}. The *redshift* therefore would be $4.2 \times 10^{-4} > \Delta v/v > 4.2 \times 10^{-8}$.

Dr. ten Bruggencate *has, in fact, been able to establish a relation between the redshift and the distribution of matter in space.* He finds $\Delta v/v \sim 1/1,000$ *for light traveling through a distance of 15,000 parsecs in the galactic plane.* It would be very important to measure the radial velocities of as many globular clusters as possible in order to decide definitely between the different theories. It is especially desirable to determine the *redshift* independent of the proper velocities of the objects observed. This might, for instance, be done with help of the steady calcium lines. It is easy to see that the above *redshift* should broaden these absorption lines asymmetrically toward the red. If these lines can be photographed with a high enough dispersion, the displacement of the center of gravity of the line will give the *redshift* independent of the velocity-of-the system from which the light is emitted.

The explanation of the apparent velocity of recession of distant nebulae proposed in this paper is in qualitative accordance with all of the observational facts known so far. It is therefore desirable, in the first place, to place the computations on a sound theoretical basis involving the *general theory of relativity*. In the second place, the transfer of *momentum* from the light to the surrounding *masses* should be determined taking into account all of the *mutual gravitational interactions*. Thirdly, it is evident that the proper motions of these *masses* will play some role. Shifts of the spectral lines to the violet should indeed be expected for thermodynamic reasons if light is traveling through systems of masses with very high average velocities. Finally, *it might be interesting to study the gravitational drag exerted by light upon light.*

I wish to thank Dr. ten Bruggencate who kindly set out on the difficult task of testing some of the suggestions presented in this paper.

Fritz Zwicky (February 14, 1898 – February 8, 1974).

Zwicky was a Swiss astronomer. He worked most of his life at the California Institute of Technology in the United States of America, where he made many important contributions in theoretical and observational astronomy. In 1933, Zwicky was the first to use the virial theorem to postulate the existence of unseen dark matter, describing it as "dunkle Materie".

Fritz Zwicky was born in Varna, Bulgaria, to a Swiss father (citizenship in Mollis, Glarus) and Czech mother. His father, Fridolin (b. 1868), was a prominent industrialist in the Bulgarian city and also served as Norwegian consul in Varna (1908–1933). Fridolin Zwicky designed and built his family's Zwicky House in Varna. Fritz's mother, Franziska Vrček (b. 1871), was an ethnic Czech of the Austro-Hungarian Empire. Fritz was the oldest of three children: he had a younger brother named Rudolf and a sister, Leonie. Fritz's mother died in Varna in 1927. His father lived and worked in Bulgaria until 1945, and returned to Switzerland after World War II. Fritz's sister Leonie married a Bulgarian from Varna and spent her entire life in the city.

In 1904, at the age of six, Fritz was sent to his paternal grandparents to Glarus, Switzerland, to study commerce. His interests shifted to math and physics. He received an advanced education in mathematics and experimental physics at the Swiss Federal Polytechnic (today known as ETH Zurich) in Zürich. He finished his studies there in 1922 with a Dr. sc. nat. degree (PhD equivalent) with a thesis entitled "Zur Theorie der heteropolaren Kristalle" (On the theory of heteropolar crystals).

In 1925, Zwicky emigrated to the United States to work with Robert Millikan at California Institute of Technology (Caltech) after receiving the Rockefeller Foundation fellowship. He had an office down the hall from Robert Oppenheimer.

On 25 March 1932, Fritz Zwicky married Dorothy Vernon Gates (1904–1991), a member of a prominent local family and a daughter of California

State Senator Egbert James Gates. Her money was instrumental in the funding of the Palomar Observatory during the Great Depression. Nicholas Roosevelt, cousin of President Theodore Roosevelt, was his brother-in-law by marriage to Tirzah Gates. Zwicky and Dorothy divorced amicably in 1941.

His first scientific contributions pertained to ionic crystals and electrolytes.

Zwicky developed numerous cosmological theories that have had a profound influence on the understanding of our universe in the early 21st century. He coined the term "supernova" while fostering the concept of neutron stars. Five years passed before Oppenheimer published his landmark paper announcing "neutron stars".

While examining the Coma galaxy cluster in 1933, Zwicky was the first to use the virial theorem to discover the existence of a gravitational anomaly, which he termed dunkle Materie '*dark matter*'. The gravitational anomaly surfaced due to the excessive rotational velocity of luminous matter compared to the calculated gravitational attraction within the cluster. He calculated the gravitational mass of the galaxies within the cluster from the observed rotational velocities and obtained a value at least 400 times greater than expected from their luminosity. The same calculation today shows a smaller factor, based on greater values for the mass of luminous material.

Together with colleague Walter Baade, Zwicky pioneered and promoted the use of the first Schmidt telescopes used in a mountain-top observatory in 1935. In 1934 he and Baade coined the term "supernova" and hypothesized that supernovae were the transition of normal stars into neutron stars, as well as the origin of cosmic rays. This was an opinion which contributed to determining the size and age of the universe subsequently.

In support of this hypothesis, Zwicky started looking for supernovae, and found a total of 120 by himself (and one more, SN 1963J, in concert with

Paul Wild) over 52 years (SN 1921B through SN 1973K), a record which stood until 2009 when passed by Tom Boles. Zwicky did his laborious work, comparing photographic plates with the human eye, which is far more challenging and difficult than Boles accomplished using modern technology for his record.

In 1937, Zwicky posited that galaxies could act as gravitational lenses by the previously discovered Einstein effect. It was not until 1979 that this effect was confirmed by observation of the so-called "Twin Quasar" Q0957+561.

When Edwin Hubble discovered a somewhat linear relationship between the *distance* to a galaxy and its *redshift* expressed as a velocity, Zwicky immediately pointed out that the correlation between the calculated distances of galaxies and their redshifts had a discrepancy too large to fit in the distance's error margins. *He proposed that the reddening effect was not due to motions of the galaxy, but to an unknown phenomenon that caused photons to lose energy as they traveled through space.* [Zwicky, F. (1929). On the Red Shift of Spectral Lines through Interstellar Space. *Proc. Natl. Acad. Sci.*, 15, 773-9; https://www.pnas.org/doi/pdf/10.1073/pnas.15.10.773.]

He considered the most likely candidate process to be a drag effect in which *photons* transfer *momentum* to surrounding masses through gravitational interactions; and proposed that an attempt be made to put this effect on a sound theoretical footing with general relativity. He rejected explanations involving the expansion of space.

Zwicky was appointed Professor of Astronomy at Caltech in 1942. He also worked as a research director/consultant for Aerojet Engineering Corporation (1943–1961), and as a staff member of Mount Wilson Observatory and Palomar Observatory for most of his career. He developed some of the earliest jet engines and holds more than 50 patents, many in jet propulsion. He invented the Underwater Jet. An example of his WWII rocket propulsion work would be a patent on a nitromethane engine filed by a collective of Zwicky and three other

Aerojet employees in March 1944, and he also published an article on chemical kinetics in rocket engines and motors in 1950.

On 18 October 1947 Zwicky married in Switzerland to Anna Margaritha Zürcher. They had three daughters together, Margrit, Franziska, and Barbarina.

Zwicky devoted considerable time to the search for galaxies and the production of catalogs. From 1961 to 1968 he and his colleagues published a comprehensive six volume Catalogue of galaxies and of clusters of galaxies. They were all published in Pasadena, by the California Institute of Technology.

Galaxies in the original catalog are called Zwicky galaxies, and the catalog is still maintained and updated today. Zwicky with his wife Margaritha also produced an important catalog of compact galaxies, sometimes called simply The Red Book.

Zwicky was an original thinker, and his contemporaries frequently had no way of knowing which of his ideas would work out and which would not. In a retrospective look at Zwicky's life and work, Stephen Maurer said:

> "When researchers talk about neutron stars, dark matter, and gravitational lenses, they all start the same way: "Zwicky noticed this problem in the 1930s. Back then, nobody listened...".""

He is celebrated for the discovery of neutron stars. In 1972, Zwicky was awarded the Gold Medal of the Royal Astronomical Society, their most prestigious award, for "distinguished contributions to astronomy and cosmology". This award noted in particular his work on neutron stars, dark matter, and cataloging of galaxies.

Zwicky produced hundreds of publications over a long career, covering a great breadth of topics.

Zwicky died in Pasadena, California on February 8, 1974, and was buried in Mollis, Switzerland.

The Fritz Zwicky Stiftung (Foundation) was established in Switzerland to carry on his ideas relating to "Morphological analysis". The foundation published a biography of Zwicky in English: Alfred Stöckli & Roland Müller: Fritz Zwicky – An Extraordinary Astrophysicist. Cambridge: Cambridge Scientific Publishers, 2011. The Zwicky Museum at the Landesbibliothek, Glarus, houses many of his papers and scientific works.

Zwicky was critical of religion and considered it unacceptable to attribute natural phenomena to God.

Interactions of Photons with Matter

MIT OpenCourseWare; https://ocw.mit.edu/courses/22-55j-principles-of-radiation-interactions-fall-2004/91d07afcf726613b748e07aebb4f1eb5_ener_depo_photon.pdf.

- *Photons* are electromagnetic radiation with zero mass, *zero charge*, and a velocity that is always c, the speed of light.
- Because they are electrically neutral, they *do not steadily lose energy* via
coulombic interactions with atomic *electrons*, as do charged particles.
- Photons travel some considerable distance before undergoing a more *"catastrophic"* interaction leading to *partial or total transfer* of the *photon* energy to *electron* energy.
- These *electrons* will ultimately deposit their energy in the medium.
- Photons are far *more penetrating* than charged particles of similar energy.

Energy Loss Mechanisms

- *photoelectric effect;*
- *Compton scattering;*
- *pair production.*

Interaction probability

- *linear attenuation coefficient, µ,*
The probability of an interaction per unit distance traveled
Dimensions of inverse length (e.g. cm^{-1}).
$$N = N_0 e^{-\mu x}.$$

- The coefficient μ depends on *photon energy* and on the *material being*
traversed.

- ***mass attenuation coefficient, μ/ρ***
The probability of an interaction per g cm^{-2} of material traversed.
Units of cm^2 g^{-1}

$$N = N_0 e^{-(\mu/\rho)(\rho x)}.$$

Mechanisms of Energy Loss: Photoelectric Effect

- In the *photoelectric absorption process*, a *photon* undergoes an interaction
with an absorber atom in which the *photon* completely disappears.
- In its place, an energetic *photoelectron* is ejected from one of the bound shells of the atom.
- For *gamma rays* of sufficient energy, the most probable origin of the
photoelectron is the most tightly bound or K shell of the atom.
- The *photoelectron* appears with an *energy* given by
$$E_{e^-} = h\nu - E_b$$
(E_b represents the *binding energy* of the *photoelectron* in its original shell).

Thus, for gamma-ray energies of more than a few hundred keV, the *photoelectron* carries off the majority of the original photon energy.

Filling of the inner shell vacancy can produce *fluorescence radiation*, or *x ray photon*(s).

The *photoelectric* process is the *predominant mode of photon interaction* at
>o *relatively low photon energies*
>o *high atomic number Z.*

The probability of *photoelectric absorption*, symbolized τ (tau), is roughly proportional to

$$\tau \propto Z^n/(h\nu)^3$$

where the exponent n varies between 3 and 4 over the gamma-ray energy region of interest.

This severe dependence of the *photoelectric absorption probability* on the atomic number of the absorber is a primary reason for the preponderance of *high-Z materials (such as lead) in gamma-ray shields.*

The photoelectric interaction is most likely to occur if the energy of the incident *photon* is *just greater than the binding energy* of the *electron* with which it interacts.

Compton Scattering

>• *Compton scattering* takes place between the incident *gamma-ray photon* and an *electron* in the absorbing material.
>• It is most often the predominant interaction mechanism for *gamma-ray* energies typical of radioisotope sources.
>• It is the *most dominant interaction mechanism in tissue.*

In Compton scattering, the incoming *gamma-ray photon* is *deflected* through an angle θ with respect to its original direction.

The *photon* transfers a portion of its *energy* to the *electron* (assumed to be initially at rest), which is then known as a *recoil electron*, or a *Compton electron.*

- All angles of scattering are possible.
- The *energy* transferred to the *electron* can vary from zero to a large fraction of the *gamma-ray energy*.
- The *Compton process* is most important for *energy absorption* for soft tissues in the range from 100 keV to 10MeV.
- The *Compton scattering* probability is symbolized σ (sigma):
    - almost *independent of atomic number Z*;
    - decreases as the *photon energy* increases;
    - directly proportional to the number of *electrons* per gram, which only varies by 20% from the lightest to the heaviest elements (except for hydrogen).

Compton Scattering Energetics

The energies of the scattered *photon* hv' and the *Compton electron* E_e, are given by

$$hv' = hv \; 1/\{1 + \alpha(1 - \cos \theta)\}$$
$$E_e = hv \; \alpha(1 - \cos \theta)/\{1 + \alpha(1 - \cos \theta)\}$$

where $\alpha = hv/m_0c^2$.

[m_0c^2 is the electron rest energy, 0.511 MeV, hv is the incoming *photon energy*.]

Limits of Energy Loss

Maximum energy transfer to recoil electron:
- angle of electron recoil is forward at 0°, $\varphi = 0°$,
- the scattered photon will be scattered straight back, $\theta = 180°$
- With $\theta = 180°$, $\cos \theta = -1$ the expressions above simplify to:

$$E_{e(max)} = hv\ 2\alpha/(1 + 2\alpha)$$

and

$$hv'_{min} = hv\ 1/(1 + 2\alpha).$$

The Table below illustrates how the amount of *energy* transferred to the *electron* varies with *photon energy*. Energy transfer is not large until the incident *photon* is in excess of approximately 100 keV.

Photon Energy, 5.11 keV	Photon Energy, 5.11 MeV
$\alpha = \ldots = 0.010$	$\alpha = \ldots = 10$
$E_{e(max)} = \ldots = 0.10$ keV	$E_{e(max)} = \ldots = 4.87$ MeV
$hv'_{min} = \ldots = 5.01$ keV	$hv'_{min} = \ldots = 0.24$ MeV
Energy transferred: 2%.	Energy transferred: 95%.

For *low-energy photons*, when the scattering interaction takes place, *little energy is transferred*, regardless of the probability of such an interaction.

As the energy increases, the fractional transfer increases, approaching 1.0 for *photons* at energies above 10 to 20 MeV.

Pair Production

If a *photon* enters matter with an energy *in excess of 1.022 MeV*, it may interact by a process called *pair production*.

The *photon*, passing near the *nucleus* of an atom, is subjected to strong field effects from the *nucleus* and may disappear as a *photon* and reappear as a positive and negative *electron pair*.

The two *electrons* produced, e^- and e^+, are not scattered *orbital electrons*, but are created, *de novo*, in the *energy/mass conversion* of the disappearing *photon*.

Pair Production Energetics

The *kinetic energy* of the *electrons* produced will be the difference between the *energy* of the incoming *photon* and the energy equivalent of two *electron* masses (2 x 0.511, or 1.022 MeV).

$$E_{e+} + E_{e-} = h\nu - 1.022 \text{ (MeV)}$$

Pair production probability, symbolized κ (kappa),
- Increases with *increasing photon energy*;
- Increases with atomic number approximately as Z^2.

Summary

- *Photoelectric effect*: produces a *scattered photon* and an *electron*, varies as $\sim Z^4/E^3$;
- *Compton effect*: produces an *electron*, varies as $\sim Z$;
- *Pair production*: produces an *electron* and a *positron*, varies as $\sim Z^2$.

Bulk Behavior of Photons in an Absorber

Attenuation Coefficients

Linear attenuation coefficient μ:
The probability of an interaction per unit distance traveled, μ, has the dimensions of inverse length (eg. cm^{-1}).

$$N = N_0 e^{-\mu x}.$$

The coefficient μ depends on *photon energy* and on the *material being traversed*.

The *mass attenuation coefficient*, μ /ρ, is obtained by dividing μ by the *density* ρ of the material, usually expressed in $cm^2 g^{-1}$.

Compton, A. H. (May, 1923) A Quantum Theory of the Scattering of X-rays by Light Elements.

Phys. Rev., 21, 5, 483-502; https://doi.org/10.1103/PhysRev.21.483.

Received December 13, 1922.

Washington University, Saint Louis.

A review of Compton's paper confirms that the scattering of *electrons* by *X-rays* or *γ-rays* results in a *redshift* due to the transfer of *energy* from the *photons* to the scattered *electron*. For *electromagnetic radiation* from remote *galaxies* observed from the Earth, travelling through *intergalactic* space known to contain electrons and other matter, this results in a *linear increase in the redshift with the distance travelled by the photons*. This contributes to the *cosmological redshift of light*, which is currently assumed to be due to the *Doppler effect* on light resulting from the *expansion of the universe* following the *Big Bang*. It is possible that both phenomena contribute to the observed *redshift* but if the *loss of energy by photons* is sufficient, this would suggest that *there was no Big Bang, the universe is not just 13.8 billion years old*, but *is indefinitely old, and in a steady state, not expanding*.

Compton's paper notes that classical electrodynamics predicts that the energy scattered by an *electron* traversed by an *X-ray* beam is independent of the *wave-length* of the incident rays. It also predicts that when the *X-rays* traverse a thin layer of matter the intensity of the scattered radiation on the two sides of the layer should be the same. But experiments on the scattering of *X-rays* by light elements showed that these predictions were correct when *X-rays* of moderate hardness were employed, but when very hard *X-rays* or *γ-rays* were employed, the scattered energy was less than Thomson's theoretical value and was strongly concentrated on the emergent side of the scattering plate. Compton applied Einstein's hypothesis to the scattering of *X-ray* and *γ-ray photons* by *electrons* and derived the mathematical relationship between the shift in *wavelength* and the scattering angle of the *X-rays, by* assuming that each scattered X-ray photon interacts with only one electron. This agreed with experimental results for the scattering of X-ray and γ-ray photons by electrons, subsequently known as *Compton scattering*, providing important evidence for *quantum theory*. The introduction of *special relativity* was irrelevant to the comparison of the theory with experimental results.

––––––––––

Abstract

A quantum theory of the scattering of X-rays and γ-rays by light elements.

The hypothesis is suggested that when an *X-ray* quantum is scattered it spends all of its *energy* and *momentum* upon some particular *electron*. This *electron* in turn scatters the ray in some definite direction. The change in *momentum* of the *X-ray* quantum due to the change in its direction of propagation results in a recoil of the scattering *electron*. The *energy* in the scattered quantum is thus less than the *energy* in the primary quantum by the *kinetic energy* of recoil of the scattering electron. The corresponding *increase in the wave-length of the scattered beam is* $\lambda_\theta - \lambda_0 = (2h/mc) \sin^2 \frac{1}{2} \theta = 0.0484 \sin^2 \frac{1}{2} \theta$, where h is the Planck constant, m is the *mass* of the scattering *electron*, c is the velocity of light, and θ is the angle between the incident and the scattered ray. Hence the increase is independent of the *wave-length*.

The distribution of the scattered radiation is found, by an indirect and not quite rigid method, to be concentrated in the forward direction according to a definite law (Eq. 27). The total energy removed from the primary beam comes out *less than that given by the classical Thomson theory* in the ratio $1/(1+2\alpha)$, where $\alpha = h/mc\lambda_0 = 0.0242/\lambda_0$. Of this energy a fraction $(1 + \alpha)/(1 + 2\alpha)$ reappears as *scattered radiation*, while the remainder is truly absorbed and transformed into *kinetic energy* of recoil of the scattering *electrons*. Hence, if σ_0 is the *scattering absorption coefficient* according to the classical theory, the coefficient according to this theory is $\sigma = \sigma_0 (1 + 2\alpha) = \sigma_s + \sigma_a$, where σ_s is the true

61

scattering coefficient $[(1 + \alpha)\sigma/(1 + 2\alpha)^2]$, and σ_a is the coefficient of absorption due to scattering $[\alpha\sigma/(1 + 2\alpha)^2]$.

Unpublished experimental results are given which show that for graphite and the Mo-K radiation the scattered radiation is longer than the primary, the observed difference $(\lambda\pi/2 - \lambda_0 = .022)$ being close to the computed value .024. In the case of scattered γ-rays, the *wave-length* has been found to vary with θ in agreement with the theory, increasing from .022 A (primary) to .068 A ($\theta = 135°$). Also, the velocity of secondary β-rays excited in light elements by γ-rays agrees with the suggestion that they are recoil *electrons*. As for the predicted variation of absorption with λ, Hewlett's results for carbon for *wave-lengths* below 0.5 A are in excellent agreement with this theory; also, the predicted concentration in the forward direction is shown to be in agreement with the experimental results, both for *X-rays* and *γ-rays*.

This remarkable *agreement between experiment and theory* indicates clearly that *scattering is a quantum phenomenon* and can be explained without introducing any new hypothesis as to the size of the *electron* or any new constants; also, that *a radiation quantum carries with it momentum as well as energy*. The restriction to light elements is due to the assumption that the constraining forces acting on the scattering *electrons* are negligible, which is probably legitimate only for the lighter elements.

The spectrum of K-rays from Mo scattered by graphite, as compared with the spectrum of the primary rays, is given in Fig. 4, showing the change of wave-length.

Radiation from a moving isotropic radiator. It is found that in a direction θ with the velocity, $I_\theta/I' = (1 - \beta)^2/(1 - \beta \cos \theta)^4 = (v_\theta/v')^4$. For the *total radiation* from a *black body* in motion to an observer at rest, $I/I' = (T/T')^4 = (v_m v_m')^4$, where the primed quantities refer to the body at rest.

J. J. Thomson's classical theory of the scattering of *X-rays*, though supported by the early experiments of Barkla and others, has been found incapable of explaining many of the more recent experiments. This theory, based upon the usual electrodynamics, leads to the result that the energy scattered by an *electron* traversed by an *X-ray* beam of unit intensity is the same whatever may be the *wave-length* of the incident rays. Moreover, when the *X-rays* traverse a thin layer of *matter*, the intensity of the scattered radiation on the two sides of the layer should be the same. Experiments on the scattering of *X-rays* by light elements have shown that these predictions are correct when *X-rays* of moderate hardness are employed; but when very hard *X-rays* or *γ-rays* are employed, the *scattered energy* is found to be decidedly less than Thomson's theoretical value, and to be *strongly concentrated on the emergent side of the scattering plate*.

Several years ago, the writer suggested that this reduced scattering of the *very short wave-length X-rays might be the result of interference between the rays scattered by different parts of the electron, if the electron's diameter is comparable with the wave-length of the radiation*. By assuming the *proper radius* for the *electron*, this hypothesis supplied a quantitative explanation of the scattering for any particular *wave-length*. But recent experiments have shown that the size of the *electron* which must thus be assumed increases with the *wave-length* of the *X-rays* employed[1], and the conception of an *electron* whose size varies with the *wave-length* of the incident rays is difficult to defend.

[1] Compton, A. H. (October, 1922). *Bull. Nat. Research Council*, 20, 10.

Recently an even more serious difficulty with the classical theory of *X-ray* scattering has appeared. It has long been known that secondary *γ-rays* are softer than the primary rays which excite them, and recent experiments have shown that this is also true of *X-rays*. By a spectroscopic examination of the secondary *X-rays* from graphite, I have, indeed, been able to show that *only a small part, if any, of the secondary X-radiation is of the same wave-length as the primary*[1].

[1] In previous papers [Compton, A. H. (1921). *Phil. Mag.*, 41, 749; (1921). *Phys. Rev.*, 18, 96] I have defended the view that the softening of the secondary X-radiation was due to a considerable admixture of a form of fluorescent radiation. Gray (1913). *Phil. Mag.*, 26, 611; (Nov, 1920). *Frank. Inst. Journ.*, p. 643) and Florance (1914). *Phil. Mag.*, 27, 225 have considered that the evidence favored true scattering, and that the softening is in some way an accompaniment of the scattering process. The considerations brought forward in the present paper indicate that the latter view is the correct one.

While the energy of the secondary *X-radiation* is so nearly equal to that calculated from Thomson s classical theory that it is difficult to attribute it to anything other than true scattering[2], these results show that if there is any scattering comparable in magnitude with that predicted by Thomson, it is of a greater *wave-length* than the primary *X-rays*.

[2] *Loc. cit.*, Compton, A. H. (Oct., 1922). *Bull. Nat. Research Council*, No. 20, p. 16.

Such a change in wave-length is directly counter to Thomson's theory of scattering, for this demands that the scattering electrons, radiating as they do because of their forced vibrations when traversed by a primary X-ray, shall give rise to radiation of exactly the same frequency as that of the radiation falling upon them. Nor does any modification of the theory such as the hypothesis of a large *electron* suggest a way out of the difficulty. This failure makes it appear improbable that a satisfactory explanation of the scattering of *X-rays* can be reached on the basis of the classical electrodynamics.

The Quantum Hypothesis of Scattering.

According to the classical theory, each *X-ray* affects every *electron* in the *matter* traversed, and the scattering observed is that due to the combined effects of all the *electrons*. *From the point of view of the quantum theory, we may suppose that any particular quantum of X-rays is not scattered by all the electrons in the radiator, but spends all of its energy upon some particular electron.* This *electron* will in turn scatter the ray in some definite direction, at an angle with the incident beam.

This bending of the path of the quantum of radiation results in a change in its *momentum*. As a consequence, *the scattering electron will recoil with a momentum equal to the change in momentum of the X-ray.* The *energy* in the scattered ray will be equal to that in the incident ray minus the *kinetic energy* of the recoil of the scattering *electron*; and since the scattered ray must be a complete quantum, the frequency will be reduced in the same ratio as is the *energy*. *Thus, on the quantum theory we should expect the wave-length of the scattered X-rays to be greater than that of the incident rays.*

> [Einstein, A. (1905). Über einen die Erzeugung und Verwandlung des Lichtes betreffenden heuristischen Gesichtspunkt. (On a Heuristic Point of View Concerning the Production and Transformation of Light): "... the energy of a *light wave* emitted from a point source is not spread continuously over ever larger volumes, but consists of a finite number of energy quanta that are spatially localized at points of space, move without dividing and are absorbed or generated only as a whole";
>
> Einstein, A. (1917). Zur Quantentheorie der Strahlung. (The Quantum Theory of Radiation): "If beam of radiation has the effect that a molecule on which it is incident absorbs or emits an amount of *energy* hv in the form of radiation by means of an elementary process, then the *momentum* hv/c is always transferred to the molecule, and, to be sure, in the case of absorption, in the direction of the moving beam and in the case of emission in the opposite direction".]

The effect of the *momentum* of the *X-ray* quantum is to set the scattering *electron* in motion at an angle of less than 90° with the primary beam. But it is well known that the energy radiated by a moving body is greater in the direction of its motion. We should therefore expect, as is experimentally observed, that *the intensity of the scattered radiation should be greater in the general direction of the primary X-rays than in the reverse direction.*

65

The change in wave-length due to scattering.

Imagine, as in Fig. 1A [original page 486], that an *X-ray* quantum of *frequency* v_0 is scattered by an *electron* of *mass* m. The *momentum* of the incident ray will be hv_0/c, where c is the velocity of light and h is Planck's constant, and that of the scattered ray is hv_θ/c at an angle θ with the initial momentum. The *principle of the conservation of momentum* accordingly demands that the *momentum* of recoil of the scattering *electron* shall equal the vector difference between the momenta of these two rays, as in Fig. 1B [original page 486]. The *momentum* of the electron, $m\beta c/\sqrt{(1-\beta^2)}$, is thus given by the relation

$$\{m\beta c/\sqrt{(1-\beta^2)}\}^2 = (hv_0/c)^2 + (hv_\theta/c)^2 + 2\ hv_0/c \cdot hv_\theta/c \cos\theta, \qquad (1)$$

where β is the ratio of the velocity of recoil of the electron to the velocity of light. But the *energy* hv_θ in the scattered quantum is equal to that of the incident quantum hv_0 less the *kinetic energy* of recoil of the scattering *electron*, i.e.,

$$hv_\theta = hv_0 - mc^2 \{1/\sqrt{(1-\beta^2)} - 1\}. \qquad (2)$$

We thus have two independent equations containing the two unknown quantities β and v_θ. On solving the equations, we find

$$v_\theta = v_0/(1 + 2\alpha \sin^2 \tfrac{1}{2}\theta), \qquad (3)$$

where

$$\alpha = hv_0/mc^2 = h/mc\lambda_0. \qquad (4)$$

Or in terms of *wave-length* instead of *frequency*,

$$\lambda_\theta = \lambda_0 + (2h/mc) \sin^2 \tfrac{1}{2}\theta. \qquad (5)$$

$$[\text{or } \lambda_\theta - \lambda_0 = h/mc\ (1 - \cos\theta)]$$

It follows from Eq. (2) that $1/(1-\beta^2) = \{1 + \alpha\ [1 - (v_\theta/v_0)]\}^2$, or solving explicitly for β

$$\beta = 2\alpha \sin \tfrac{1}{2}\theta \sqrt{\{1 + (2\alpha + \alpha^2) \sin^2 \tfrac{1}{2}\theta\}/\{1 + 2(\alpha + \alpha^2) \sin^2 \tfrac{1}{2}\theta\}} \qquad (6)$$

Eq. (5) indicates an increase in *wave-length* due to the scattering process which varies from a few per cent in the case of ordinary *X-rays* to more than 200 per cent in the case of *γ-rays* scattered backward. At the same time the *velocity* of the recoil of the scattering *electron*, as calculated from Eq. (6), varies from zero when the ray is scattered directly forward to about 80 per cent of the speed of light when a *γ-ray* is scattered at a large angle.

It is of interest to notice that *according to the classical theory*, if an *X-ray* were scattered by an *electron* moving in the direction of propagation at a *velocity* $\beta'c$, the *frequency* of the ray scattered at an angle θ is given by the *Doppler principle* as

$$\nu_\theta = \nu_0/(1 + 2\beta'/(1 - \beta') \sin^2 \tfrac{1}{2} \theta). \tag{7}$$

It will be seen that this is of exactly the same form as Eq. (3), derived on the hypothesis of the recoil of the scattering *electron*. Indeed, if $\alpha = \beta'/(1 - \beta')$ or $\beta' = \alpha/(1 + \alpha)$, the two expressions become identical. *It is clear, therefore, that so far as the effect on the wave-length is concerned, we may replace the recoiling electron by a scattering electron moving in the direction of the incident beam at a velocity such that*

$$\beta^* = \alpha/(1 + \alpha). \tag{8}$$

We shall call βc the "*effective velocity*" of the scattering *electrons*.

Energy distribution from a moving, isotropic radiator.

In preparation for the investigation of the *spatial distribution* of the *energy* scattered by a recoiling *electron*, let us study the energy radiated from a moving, isotropic body. If *an observer moving with the radiating body* draws a sphere about it, the condition of *isotropy* means that the probability is equal for all directions of emission of each *energy* quantum. That is, the probability that a quantum will traverse the sphere between the angles θ' and $\theta' + d\theta'$ with the direction of motion is $\tfrac{1}{2} \sin \theta' d\theta'$. But the surface which the *moving observer* considers a sphere (Fig. 2A [original page 488]) is considered by the *stationary observer* to be an oblate spheroid whose polar axis is reduced by the factor $\sqrt{(1 - \beta^2)}$.

Consequently, a quantum of radiation which traverses the sphere at the angle θ', whose tangent is y'/x' (Fig. 2A), appears to the stationary observer to traverse the spheroid at an angle θ'' whose tangent is y''/x'' (Fig. 2B [original page 488]). Since $x' = x''/\sqrt{(1 - \beta^2)}$ and $y' = y''$, we have

$$\tan \theta' = y'/x' = \sqrt{(1 - \beta^2)}\ y''/x'' = \sqrt{(1 - \beta^2)} \tan \theta'', \qquad (9)$$
and $\quad \sin \theta' = \{\sqrt{(1 - \beta^2)} \tan \theta''\}/\sqrt{\{1 + (1 - \beta^2) \tan^2 \theta''\}}. \qquad (10)$

...

Imagine, as in Fig. 3, that a quantum is emitted at the instant $t = 0$, when the radiating body is at O. If it traverses the *moving observer*'s sphere at an angle θ', it traverses the corresponding oblate spheroid, imagined by the stationary observer to be moving with the body, at an angle θ''. After 1 second, the quantum will have reached some point P on a sphere of radius c drawn about O, while the radiator will have moved a distance βc. The stationary observer at P therefore finds that the radiation is coming to him from the point O, at an angle θ with the direction of motion. That is, if the moving observer considers the quantum to be emitted at an angle θ' with the direction of motion, to the stationary observer the angle appears to be θ, where

$$\sin \theta /\sqrt{(1 + \beta^2 - 2\beta \cos \theta)} = \sin \theta'', \qquad (11)$$

and θ'' is given in terms of θ' by Eq. (10). It follows that

$$\sin \theta' = \sin \theta \sqrt{(1 - \beta^2)} /(1 - \beta \cos \theta). \qquad (12)$$

On differentiating Eq. (12) we obtain

$$d\theta' = \sqrt{(1 - \beta^2)} /(1 - \beta \cos \theta)\ d\theta. \qquad (13)$$

The probability that a given quantum will appear to the stationary observer to be emitted between the angles θ and $\theta + d\theta$ is therefore

$$P_\theta\ d\theta = P_{\theta'}\ d\theta' = \tfrac{1}{2} \sin \theta'\ d\theta',$$

where the values of $\sin \theta'$ and $d\theta'$ are given by Eqs. (12) and (13). Substituting these values, we find

$$P_\theta\ d\theta = (1 - \beta^2)/(1 - \beta \cos \theta)^2 \cdot \tfrac{1}{2} \sin \theta\ d\theta.$$

Suppose the *moving observer* notices that n' quanta are emitted per second. *The stationary observer will estimate the rate of emission as*

$$n'' = n' \sqrt{(1 - \beta^2)},$$

quanta per second, because of the difference in rate of the moving and stationary clocks.

Of these n" quanta, the number which are emitted between angles θ and $\theta + d\theta$ is $dn'' = n'' P_\theta d\theta$. But if dn'' per second are emitted at the angle θ, the number per second received by a stationary observer at this angle is $dn = dn''/(1 - \beta \cos \theta)$, since the radiator is approaching the observer at a velocity $\beta \cos \theta$. The energy of each quantum is, however, $h\nu_\theta$, where ν_θ is the frequency of the radiation as received by the stationary observer[1].

> [1] At first sight the assumption that the quantum which to the moving observer had energy $h\nu'$ will be $h\nu$ for the stationary observer seems inconsistent with the energy principle. When one considers, however, the work done by the moving body against the back-pressure of the radiation, it is found that the energy principle is satisfied. *The conclusion reached by the present method of calculation is in exact accord with that which would be obtained according to Lorenz's equations, by considering the radiation to consist of electromagnetic waves.*
>
> [The assumption of *Special Relativity* makes no difference to the calculation of the Compton Effect.]

Thus, the *intensity*, or the *energy* per unit area per unit time, of the radiation received at an angle θ and a distance R is

$$I_\theta = h\nu_\theta \cdot dn / 2\pi R^2 \sin \theta \, d\theta = \dots$$
$$= n'h\nu_\theta/4\pi R^2 \, (1 - \beta^2)^{3/2}/(1 - \beta \cos \theta)^3. \qquad (15)$$

If the *frequency* of the oscillator emitting the radiation is measured by an observer moving with the radiator as ν', the stationary observer judges its frequency to be $\nu'' = \nu'\sqrt{(1 - \beta^2)}$, and, in virtue of the *Doppler effect*, the *frequency* of the radiation received at an angle θ is

$$v_\theta = v''/(1 - \beta \cos \theta) = v' \, [\sqrt{(1 - \beta^2)}/ (1 - \beta \cos \theta)]. \qquad (16)$$

Substituting this value of v_θ in Eq. (15) we find

$$I_\theta = n'hv'/4\pi R^2 \, (1 - \beta^2)^2/(1 - \beta \cos \theta)^4. \qquad (17)$$

But the *intensity* of the radiation observed by the moving observer at a distance R from the source is $I' = n'hv'/4\pi R^2$. Thus,

$$I_\theta = I' \, [(1 - \beta^2)^2/(1 - \beta \cos \theta)^4] \qquad (18)$$

is the *intensity* of the radiation received at an angle θ with the direction of motion of an *isotropic* radiator, which moves with a *velocity* βc, and which would radiate with *intensity* I' if it were at rest[2].

[2] Livens, G. H. (1918). *The Theory of Electricity*, p. 6oo, gives for I_θ/I' the value $(1 - \beta \cos \theta)^{-2}$. At small velocities this value differs from the one here obtained by the factor $(1 - \beta \cos \theta)^{-2}$. The difference is due to Livens' neglect of the concentration of the radiation in the small angles, as expressed by our Eq. (14). Cunningham (1914). *The Principle of Relativity*, p. 60, shows that if a plane wave is emitted by a radiator moving in the direction of propagation with a velocity βc, the intensity I received by a stationary observer is greater than the *intensity* I' estimated by the moving observer, in the ratio $(1 - \beta^2)/(1 - \beta)^2$, which is in accord with the value calculated according to the methods here employed. The *change in frequency* given in Eq. (16) is that of the usual *relativity theory*. I have not noticed the publication of any result which is the equivalent of my formula (18) for the *intensity of the radiation* from a moving body.

It is interesting to note, on comparing Eqs. (16)

$$[v_\theta = v' \, [\sqrt{(1 - \beta^2)}/ (1 - \beta \cos \theta)] \qquad (16)]$$

and (18)

$$[I_\theta = I' \, [(1 - \beta^2)^2/(1 - \beta \cos \theta)^4], \qquad (18)]$$

that

$$I_\theta/I' = (v_\theta/v')^4. \qquad (19)$$

This result may be obtained very simply for the *total radiation* from a *black body*, which is a special case of an *isotropic* radiator. For, suppose such a radiator is moving so that the *frequency* of maximum intensity which to a moving observer is I ' appears to the stationary observer to be v_m'. Then according to *Wien's law*, the *apparent temperature* T, as estimated by the stationary observer, is greater than the *temperature* T' for the moving observer by the ratio $T/T' = v_m/v_m'$. According to *Stefan's law*, however, the intensity of the *total radiation* from a *black body* is proportional to T^4, hence, if I and I' are the *intensities* of the radiation as measured by the stationary and the moving observers respectively,

$$I_\theta/I' = (T/T')^4 = (v_m/v_m')^4. \tag{20}$$

The agreement of this result with Eq. (19) may be taken as confirming the correctness of the latter expression.

The intensity of scattering from recoiling electrons.

We have seen that *the change in frequency of the radiation scattered by the recoiling electrons is the same as if the radiation were scattered by electrons moving in the direction of propagation with an effective velocity $\beta^* = \alpha/(1 + \alpha)$, [Eq. (8)] where $\alpha = h/mc\lambda_0$ [Eq. 4].* It seems obvious that since these two methods of calculation result in the same change in wave-length, they must *also result in the same change in intensity of the scattered beam.* This assumption is supported by the fact that we find, as in Eq. 19, that the change in intensity is in certain special cases a function only of the change in *frequency.* I have not, however, succeeded in showing rigidly that if two methods of scattering result in the same relative *wave-lengths* at different angles, they will also result in the same relative intensity at different angles. *Nevertheless, we shall assume that this proposition is true,* and shall proceed to calculate the relative intensity of the scattered beam at different angles on the hypothesis that the scattering electrons are moving in the direction of the primary beam with a *velocity $\beta^* = \alpha/(1 + \alpha)$.* If our assumption is correct, the results of the calculation will apply also to the scattering by recoiling *electrons.*

To an observer moving with the scattering *electron*, the intensity of the scattering at an angle θ', according to the usual electrodynamics, should be proportional to $(1 + \cos^2 \theta')$, if the primary beam is unpolarized. On the quantum theory, this means that the probability that a quantum will be emitted between the angles θ' and $\theta' + d\theta'$ is proportional to $(1 + \cos^2 \theta').\sin \theta'd\theta'$; since $2\pi \sin \theta'd\theta'$ is the solid angle included between θ' and $\theta' + d\theta'$. This may be written $P_{\theta'} d\theta' = k(1 + \cos^2 \theta') \sin \theta'd\theta'$.

The factor of proportionality k may be determined by performing the integration

$$\int_0^\pi P_{\theta'} d\theta' = k \int_0^\pi (1 + \cos^2 \theta') \sin \theta'd\theta' = 1,$$

with the result that $k = 3/8$. Thus

$$P_{\theta'} d\theta' = 3/8 (1 + \cos^2 \theta') \sin \theta'd\theta' \tag{21}$$

is the probability that a quantum will be emitted at the angle θ' *as measured by an observer moving with the scattering electron.*

To the stationary observer, however, the quantum ejected at an angle θ' appears to move at an angle θ with the direction of the primary beam, where $\sin \theta'$ and $d\theta'$ are as given in Eqs. (12) and (13). Substituting these values in Eq. (21), we find for the probability that a given quantum will be scattered between the angles θ and $\theta + d\theta$,

$$P_{\theta'} d\theta' = 3/8 \sin\theta \, d\theta \, [(1 - \beta^2)\{(1 + \beta^2)(1 + \cos^2\theta) - 4\beta \cos\theta\}]/(1 - \beta \cos\theta)^4 \tag{22}$$

Suppose the stationary observer notices that n quanta are scattered per second. In the case of the radiator emitting n" quanta per second while approaching the observer, the n"th quantum was emitted when the radiator was nearer the observer, so that the interval between the receipt of the 1st and the n"th quantum was less than a second. That is, more quanta were received per second than were emitted in the same time. In the case of scattering, however, though we suppose that each scattering *electron* is moving forward, the nth quantum is scattered by an *electron* starting from the same position as the 1st quantum. Thus, the number of quanta received per second is also n.

We have seen (Eq. 3) that the *frequency* of the quantum received at an angle θ is

$$\nu_\theta = \nu_0/(1 + 2\alpha \sin^2 \tfrac{1}{2} \theta) = \nu_0/\{1 + \alpha (1 - \cos \theta)\},$$

where ν_0, the *frequency* of the incident beam, is also the *frequency* of the ray scattered in the direction of the incident beam. The *energy* scattered per second at the angle θ is thus $nh\nu_\theta P_\theta d\theta$, and the *intensity*, or *energy* per second per unit area, of the ray scattered to a distance R is

$$I_\theta = nh\nu_\theta P_\theta d\theta/2\pi R^2 \sin \theta d\theta = \dots .$$

Substituting for β its value $\alpha/(1 + \alpha)$, and reducing, this becomes

$$I = \dots .$$

In the forward direction, where $\theta = 0$, the *intensity* of the scattered beam is thus

$$I_0 = 3/8\pi \ nh\nu_0/R^2 (1 + 2\alpha). \tag{24}$$

Hence

$$I_\theta/I_0 = \tfrac{1}{2} \{1 + \cos^2 \theta + 2\alpha (1 + \alpha)(1 - \cos \theta)^2\}/\{1 + \alpha(1 - \cos \theta)\}^5 \tag{25}$$

On the hypothesis of recoiling *electrons*, however, for a ray scattered directly forward, the *velocity* of recoil is zero (Eq. 6). Since in this case the scattering *electron* is at rest, *the intensity of the scattered beam should be that calculated on the basis of the classical theory*, namely,

$$I_0 = I (Ne^4/R^2 m^2 c^4), \tag{26}$$

where I is the *intensity* of the primary beam traversing the N *electrons* which are effective in scattering. On combining this result with Eq. (25), we find for the *intensity* of the X-rays scattered at an angle θ with the incident beam,

$$I_\theta = I \ Ne^4/2R^2 m^2 c^4 \{1 + \cos^2\theta + 2\alpha (1 + \alpha)(1 - \cos\theta)^2\}$$
$$/\{1 + \alpha(1 - \cos \theta)\}^5. \tag{27}$$

The calculation of the *energy* removed from the primary beam may now be made without difficulty. We have supposed that n quanta are scattered per second. But on comparing Eqs. (24) and (26), we find that

$$n = 8\pi/3 \ INe^4/h\nu_0 m^2 c^4(1 + 2\alpha).$$

The *energy* removed from the primary beam per second is $nh\nu_0$. If we define *the scattering absorption coefficient* as the fraction of the *energy* of the primary beam removed by the scattering process per unit length of path through the medium, it has the value

$$\sigma = nh\nu_0/I = 8\pi/3 \; Ne^4/m^2c^4 \; 1/(1 + 2\alpha) = \sigma_0/(1 + 2\alpha), \qquad (28)$$

where N is the number of scattering *electrons* per unit volume, and σ_0 is the scattering coefficient calculated on the basis of the classical theory[1].

[1] Cf. Thomson, J. J. *Conduction of Electricity through Gases*, 2d ed., p. 325.

In order to determine the *total energy* truly scattered, we must integrate the scattered *intensity* over the surface of a sphere surrounding the scattering material, i.e.,

$$\varepsilon_s = \int_0^\pi I_\theta . 2\pi R^2 \sin\theta \; d\theta.$$

On substituting the value of I_θ from Eq. (27), and integrating, this becomes

$$\varepsilon_s = 8\pi/3 \; INe^4/m^2c^4 \; (1 + \alpha)/(1 + 2\alpha)^2.$$

The *true scattering coefficient* is thus

$$\sigma_s = 8\pi/3 \; Ne^4/m^2c^4 \; (1 + \alpha)/(1 + 2\alpha)^2 = \sigma_0 \; (1 + \alpha)/(1 + 2\alpha)^2. \qquad (29)$$

It is clear that the difference between the *total energy* removed from the primary beam and that which reappears as scattered radiation is the *energy* of recoil of the scattering *electrons*. This difference represents, therefore, a type of true *absorption* resulting from the scattering process. The corresponding *coefficient of true absorption due to scattering* is

$$\sigma_a = \sigma - \sigma_s$$
$$= 8\pi/3 \; Ne^4/m^2c^4 \; \alpha/(1 + 2\alpha)^2 = \sigma_0 \; \alpha/(1 + 2\alpha)^2. \qquad (30)$$

EXPERIMENTAL TEST.

Let us now investigate the agreement of these various formulas with experiments on the change of *wave-length* due to scattering, and on the magnitude of the scattering of *X-rays* and γ-rays by light elements.

Wave-length of the scattered rays.

If in Eq. (5)

$$[\lambda_\theta = \lambda_0 + (2h/mc) \sin^2 \tfrac{1}{2}\, \theta. \tag{5}]$$

we substitute the accepted values of h, m, and c, we obtain

$$\lambda_\theta = \lambda_0 + 0.0484 \sin^2 \tfrac{1}{2}\, \theta, \tag{31}$$

if λ is expressed in Angstrom units. *It is perhaps surprising that the increase should be the same for all wave-lengths.* Yet, as a result of an extensive experimental study of the change in *wave-length* on scattering, the writer has concluded that "over the range of primary rays from 0.7 to 0.025 A, the *wave-length* of the secondary *X-rays* at 90° with the incident beam is roughly 0.03 A greater than that of the primary beam which excites it."[1]

[1] Compton, A. H. (1922). *Bull. N. R. C.*, No. 20, p. 17.

Thus, the experiments support the theory in showing a *wave-length* increase which seems independent of the incident *wavelength*, and which also is of the proper order of magnitude.

A quantitative test of the accuracy of Eq. (31)

$$[\lambda_\theta = \lambda_0 + 0.0484 \sin^2 \tfrac{1}{2}\, \theta, \tag{31}]$$

is possible in the case of the characteristic K-rays from molybdenum when scattered by graphite. In Fig. 4 [on original page 495] is shown a spectrum of the *X-rays* scattered by graphite at right angles with the primary beam, when the graphite is traversed by *X-rays* from a molybdenum target[2].

[2] It is hoped to publish soon a description of the experiments on which this figure is based.

The solid line represents the spectrum of these scattered rays, and is to be compared with the broken line, which represents the spectrum of the primary rays, using the same slits and crystal, and the same potential on the tube. The primary spectrum is, of course, plotted on a much smaller scale than the secondary. The zero point for the spectrum of both the primary and secondary *X-rays* was determined by finding the position of the first order lines on both sides of the zero point.

...

It will be seen that the *wave-length* of the scattered rays is unquestionably greater than that of the primary rays which excite them. Thus, the Kα line from molybdenum has a wave-length 0.708 A. The *wave-length* of this line in the scattered beam is found in these experiments, however, to be 0.730 A. That is,

$$\lambda_\theta - \lambda_0 = 0.022 \text{ A (experiment).}$$

But according to the present theory (Eq. 5),
$$[\lambda_\theta = \lambda_0 + (2h/mc) \sin^2 \tfrac{1}{2}\,\theta. \tag{5)]}$$

$$\lambda_\theta - \lambda_0 = 0.0484 \sin^2 45^\circ = 0.024 \text{ A (theory),}$$

which is a very satisfactory agreement.

...

There is thus good reason for believing that Eq. (5)
$$[\lambda_\theta = \lambda_0 + (2h/mc) \sin^2 \tfrac{1}{2}\,\theta. \tag{5)]}$$
represents accurately the *wave-length* of the *X-rays* and γ-rays scattered by light elements.

Velocity of recoil of the scattering electrons.

The *electrons* which recoil in the process of the scattering of ordinary *X-rays* have not been observed. This is probably because their *number* and *velocity* are usually small compared with the *number* and *velocity* of the *photoelectrons* ejected as a result of the characteristic fluorescent absorption. I have pointed out elsewhere[1], however,

[1] Compton, A. H. (1922). *Bull. N. R. C.*, No. 20, p. 27.

that there is good reason for believing that most of the secondary *β-rays* excited in light elements by the action of *γ-rays* are such recoil *electrons*. According to Eq. (6),

$$[\beta = 2\alpha \sin \tfrac{1}{2}\theta \sqrt{\{1 + (2\alpha + \alpha^2) \sin^2 \tfrac{1}{2}\theta\}/\{1 + 2(\alpha + \alpha^2) \sin^2 \tfrac{1}{2}\theta\}} \quad (6)]$$

the *velocity* of these *electrons* should vary from 0, when the *γ-ray* is scattered forward, to $v_{max} = \beta_{max}c = 2c\alpha[(1 + \alpha)/(1 + 2\alpha + 2\alpha^2)]$, when the *γ-ray* quantum is scattered backward. If for the hard *γ-rays* from radium C, $\alpha = 1.09$, corresponding to $\lambda = 0.022$ A, we thus obtain $\beta_{max} = 0.82$. The *effective velocity* of the scattering electrons is, therefore (Eq. 8),

$$[\beta^* = \alpha/(1 + \alpha). \qquad (8)]$$

$\beta^* = 0.52$. These results are in accord with the fact that *the average velocity of the β-rays excited by the γ-rays from radium is somewhat greater than half that of light*[2].

[2] Rutherford, E. *Radioactive Substances and their Radiations*, p. 273.

[β-rays are fast moving *electrons*; *γ-rays* and *X-rays* are *electromagnetic radiation* with short *wavelength* that moves with the speed of light.]

Absorption of X-rays due to scattering.

Valuable information concerning the magnitude of the scattering is given by the measurements of the *absorption* of *X-rays* due to scattering. Over a wide range of *wavelengths*, the formula for the *total mass absorption*, $\mu/\rho = \kappa \lambda^3 + \sigma/\rho$, is found to hold, where μ is the linear *absorption coefficient*, ρ is the *density*, κ is a constant, and σ is the *energy loss* due to the scattering process. ...

...

True absorption due to scattering has not been noticed in the case of *X-rays*. In the case of hard *γ-rays*, however, Ishino has shown[1] that there is *true absorption* as well as scattering, and that for the lighter elements the *true absorption* is proportional to the atomic number.

[1] Ishino, M. (1917). *Phil, Mag.*, 33, 140.

That is, *this absorption is proportional to the number of electrons present, just as is the scattering*. He gives for the *true mass absorption coefficient* of the hard *γ-rays* from RaC in both aluminum and iron the value 0.021. According to Eq. (30), the *true mass absorption* by aluminum should be 0.021 and by iron, 0.020, taking the effective *wavelength* of the rays to be 0.022 A. *The difference between the theory and the experiments is less than the probable experimental error.*

Ishino has also estimated the *true mass scattering coefficients* of the hard *γ-rays* from RaC by aluminum and iron to be 0.045 and 0.042 respectively[2].

[2] Ishino, M., *loc. cit.*

These values are very far from the values 0.193 and 0.187 predicted by the classical theory. But taking $\lambda = 0.022$ A, as before, the corresponding values calculated from Eq. (29) are 0.040 and 0.038, *which do not differ seriously from the experimental values.*

It is well known that for soft *X-rays* scattered by light elements the *total scattering* is in accord with Thomson's formula. This is in agreement with the present theory, according to which the *true scattering coefficient* σ_s, approaches Thomson's value σ_0 when $\alpha = h/mc\lambda$ becomes small (Eq. 29)
$$[\sigma_s = 8\pi/3\ Ne^4/m^2c^4\ (1 + \alpha)/(1 + 2\alpha)^2 = \sigma_0\ (1 + \alpha)/(1 + 2\alpha)^2.\ (29)]$$

The relative intensity of the X-rays scattered in different directions with the primary beam.

Our Eq. (27)
$$[I_\theta = I\ Ne^4/2R^2m^2c^4\ \{1 + \cos^2\theta + 2\alpha\ (1 + \alpha)(1 - \cos\theta)^2\}$$
$$/\{1 + \alpha(1 - \cos\theta)\}^5.\ \quad\quad\quad (27)]$$
predicts a concentration of the energy in the forward direction. A large number of experiments on the scattering of *X-rays* have shown that, except for the excess scattering at small angles, the ionization due to the scattered beam is symmetrical on the emergence and incidence sides of a scattering plate. The difference in *intensity* on the two sides according to Eq. (27) should, however, be noticeable. Thus, if the *wave-length* is

78

0.7 A, which is probably about that used by Barkla and Ayers in their experiments on the scattering by carbon[1],

[1] *Loc. cit.*, Barkla & Ayers, (1911). *Phil. Mag.*, 21, 275.

the ratio of the *intensity* of the rays scattered at 40° to that at 140° should be about 1.10. But their experimental ratio was 1.04, which differs from our theory by more than their probable experimental error.

It will be remembered, however, that our theory, and experiment also, indicates a difference in the *wave-length* of the *X-rays* scattered in different directions. The softer *X-rays* which are scattered backward are the more easily absorbed and, though of smaller *intensity*, may produce an *ionization* equal to that of the beam scattered forward. Indeed, if α is small compared with unity, as is the case for ordinary *X-rays*, Eq. (27) may be written approximately $I_\theta/I_\theta' = (\lambda_0/\lambda_\theta)^3$, where I_θ' is the *intensity* of the beam scattered at the angle θ according to the classical theory. The part of the *absorption* which results in the *ionization* is however proportional to λ^3. Hence if, as is usually the case, only a small part of the X-rays entering the ionization chamber is *absorbed* by the gas in the chamber, the *ionization* is also proportional to λ^3. Thus, if i_θ represents the *ionization* due to the beam scattered at the angle θ, and if i_θ' is the corresponding *ionization* on the classical theory, we have $i_\theta/i_\theta' = (I_\theta/I_\theta')(\lambda_0/\lambda_\theta)^3 = 1$, or $i_\theta = i_\theta'$. That is, to a first approximation the *ionization* should be the same as that on the classical theory, though the *energy* in the scattered beam is less. This conclusion is in good accord with the experiments which have been performed on the scattering of ordinary *X-rays*, if correction is made for the excess scattering which appears at small angles.

...

In the case of very short *wave-lengths*, however, the case is different. The writer has measured the *γ-rays* scattered at different angles by iron, using an ionization chamber so designed as to absorb the greater part of even the primary *γ-ray* beam[1].

[1] Compton, A. H. (1921). *Phil. Mag.*, 41, 758.

...

As before, the *wave-length* of the *γ-rays* is taken as 0.022 A. The beautiful agreement between the theoretical and the experimental values of the scattering is the more striking when one notices that there is not a single adjustable constant connecting the two sets of values.

Discussion.

This remarkable agreement between our formulas and the experiments can leave but little doubt that the scattering of X-rays is a quantum phenomenon. The hypothesis of a large *electron* to explain these effects is accordingly superfluous, for all the experiments on *X-ray* scattering to which this hypothesis has been applied are now seen to be explicable from the point of view of the *quantum theory* without introducing any new hypotheses or constants. In addition, *the present theory accounts satisfactorily for the change in wave-length due to scattering*, which was left unaccounted for on the hypothesis of the large *electron*. From the standpoint of the scattering of *X-rays* and *γ-rays*, therefore, there is no longer any support for the hypothesis of an *electron* whose diameter is comparable with the *wave-length* of hard *X-rays*.

The present theory depends essentially upon the assumption that each electron which is effective in the scattering scatters a complete quantum. It involves also the hypothesis that the quanta of radiation are received from definite directions and are scattered in definite directions. The experimental support of the theory indicates very convincingly that a radiation quantum carries with it directed *momentum* as well as *energy*.

Emphasis has been laid upon the fact that in its present form the *quantum theory of scattering* applies only to light elements. The reason for this restriction is that we have tacitly assumed that there are no forces of constraint acting upon the scattering *electrons*. This assumption is probably legitimate in the case of the very light elements, but cannot be true for the heavy elements. For if the *kinetic energy* of recoil of an *electron* is less than the *energy* required to remove the *electron* from the atom, there is no chance for the *electron* to recoil in the manner we have

80

supposed. The conditions of scattering in such a case remain to be investigated.

The manner in which *interference* occurs, as for example in the cases of excess scattering and *X-ray* reflection, is not yet clear. Perhaps if an *electron* is bound in the atom too firmly to recoil, the incident quantum of radiation may spread itself over a large number of *electrons*, distributing its *energy* and *momentum* among them, thus making *interference* possible. In any case, the problem of scattering is so closely allied with those of *reflection* and *interference* that a study of the problem may very possibly shed some light upon the difficult question of the relation between *interference* and the *quantum theory*.

Many of the ideas involved in this paper have been developed in discussion with Professor G. E. M. Jauncey of this department.

Arthur Holly Compton (September 10, 1892 – March 15, 1962).

Compton was an American physicist who won the Nobel Prize in Physics in 1927 for his 1923 discovery of the *Compton effect*, which demonstrated the particle nature of electromagnetic radiation. It was a sensational discovery at the time: the wave nature of light had been well-demonstrated, but the idea that light had both wave and particle properties was not easily accepted. He is also known for his leadership over the Metallurgical Laboratory at the University of Chicago during the Manhattan Project, and served as chancellor of Washington University in St. Louis from 1945 to 1953.

Compton was born on September 10, 1892, in Wooster, Ohio, the son of Elias and Otelia Catherine (née Augspurger) Compton, who was named American Mother of the Year in 1939. They were an academic family. Elias was dean of the University of Wooster (later the College of Wooster), which Arthur also attended. Arthur's eldest brother, Karl, who also attended Wooster, earned a Doctor of Philosophy (PhD) degree in physics from Princeton University in 1912, and was president of the Massachusetts Institute of Technology from 1930 to 1948. His second brother Wilson likewise attended Wooster, earned his PhD in economics from Princeton in 1916 and was president of the State College of Washington, later Washington State University from 1944 to 1951.

Compton was initially interested in astronomy, and took a photograph of Halley's Comet in 1910. Around 1913, he described an experiment where an examination of the motion of water in a circular tube demonstrated the rotation of the earth, a device now known as the Compton generator. That year, he graduated from Wooster with a Bachelor of Science degree and entered Princeton, where he received his Master of Arts degree in 1914. Compton then studied for his PhD in physics under the supervision of Hereward L. Cooke, writing his dissertation on *The Intensity of X-Ray Reflection, and the Distribution of the Electrons in Atoms*.

When Compton earned his PhD in 1916, he, Karl and Wilson became the first group of three brothers to earn PhDs from Princeton. Later, they would become the first such trio to simultaneously head American colleges. Their sister Mary married a missionary, C. Herbert Rice, who became the principal of Forman Christian College in Lahore. In June 1916, Compton married Betty Charity McCloskey, a Wooster classmate and fellow graduate. They had two sons, Arthur Alan Compton and John Joseph Compton.

Compton spent a year as a physics instructor at the University of Minnesota in 1916–17, then two years as a research engineer with the Westinghouse Lamp Company in Pittsburgh, where he worked on the development of the sodium-vapor lamp. During World War I he developed aircraft instrumentation for the Signal Corps.

In 1919, Compton was awarded one of the first two National Research Council Fellowships that allowed students to study abroad. *He chose to go to the University of Cambridge's Cavendish Laboratory in England.* Working with George Paget Thomson, the son of J. J. Thomson, *Compton studied the scattering and absorption of gamma rays.* He observed that the scattered rays were more easily absorbed than the original source.

Compton was greatly impressed by the Cavendish scientists, especially Ernest Rutherford, Charles Galton Darwin and Arthur Eddington, and he ultimately named his second son after J. J. Thomson.

Returning to the United States, Compton was appointed Wayman Crow Professor of Physics, and head of the Department of Physics at Washington University in St. Louis in 1920. *In 1922, he found that X-ray quanta scattered by free electrons had longer wavelengths and, in accordance with Planck's relation, less energy than the incoming X-rays, the surplus energy having been transferred to the electrons. This discovery, known as the "Compton effect" or "Compton scattering", demonstrated the particle concept of electromagnetic radiation.*

In 1923, Compton published a paper in the *Physical Review* that explained the X-ray shift by attributing particle-like momentum to photons, something Einstein had invoked for his 1905 Nobel Prize–winning explanation of the photo-electric effect. First postulated by Max Planck in 1900, these were conceptualized as elements of light "quantized" by containing a specific amount of energy depending only on the frequency of the light. In his paper, Compton derived the mathematical relationship between the shift in wavelength and the scattering angle of the X-rays *by assuming that each scattered X-ray photon interacted with only one electron*. His paper concludes by reporting on experiments that verified his derived relation:

$$\lambda_\theta - \lambda_0 = h/mc \, (1 - \cos \theta)$$

where

λ_0 is the initial wavelength,

λ_θ is the wavelength after scattering,

h is the Planck constant,

m is the electron rest mass,

c is the speed of light, and

θ is the scattering angle.

[This formula is not included in his 1923 paper but can be derived from the formula that were. See paper below.]

The quantity h/mc is known as the Compton wavelength of the electron; it is equal to 2.43×10^{-12} m. The wavelength shift $\lambda_\theta - \lambda_0$ lies between zero (for $\theta = 0°$) and twice the Compton wavelength of the electron (for $\theta = 180°$). He found that some X-rays experienced no wavelength shift despite being scattered through large angles; in each of these cases the photon failed to eject an electron. In these cases, the magnitude of the shift is related not to the Compton wavelength of the electron, but to the Compton wavelength of the entire atom, which can be upwards of 10,000 times smaller.

"When I presented my results at a meeting of the American Physical Society in 1923", Compton later recalled, "it initiated the most hotly contested scientific controversy that I have ever known." The wave nature of light had been well demonstrated, and the idea that it could have a dual nature was not easily accepted. It was particularly telling that diffraction in a crystal lattice could only be explained with reference to its wave nature. It earned Compton the Nobel Prize in Physics in 1927. Compton and Alfred W. Simon developed the method for observing at the same instant individual scattered X-ray photons and the recoil electrons. In Germany, Walther Bothe and Hans Geiger independently developed a similar method.

In 1923, Compton moved to the University of Chicago as professor of physics, a position he would occupy for the next 22 years. In 1925, he demonstrated that the scattering of 130,000-volt X-rays from the first sixteen elements in the periodic table (hydrogen through sulfur) were polarized, a result predicted by J. J. Thomson. He used X-rays to investigate ferromagnetism, concluding that it was a result of the alignment of electron spins.

Compton's first book, *X-Rays and Electrons*, was published in 1926. In it he showed how to calculate the densities of diffracting materials from their X-ray diffraction patterns. He revised his book with the help of Samuel K. Allison to produce *X-Rays in Theory and Experiment* (1935). This work remained a standard reference for the next three decades.

In 1926, he became a consultant for the Lamp Department at General Electric. In 1934, he returned to England as Eastman visiting professor at Oxford University. While there, General Electric asked him to report on activities at General Electric Company plc's research laboratory at Wembley. Compton was intrigued by the possibilities of the research there into fluorescent lamps. His report prompted a research program in America that developed it.

By the early 1930s, Compton had become interested in cosmic rays. At the time, their existence was known but their origin and nature remained speculative. Their presence could be detected using a spherical "bomb" containing compressed air or argon gas and measuring its electrical conductivity. Trips to Europe, India, Mexico, Peru and Australia gave Compton the opportunity to measure cosmic rays at different altitudes and latitudes. Along with other groups who made observations around the globe, they found that *cosmic rays were 15% more intense at the poles than at the equator*. Compton attributed this to the effect of cosmic rays being made up principally of charged particles, rather than photons as Robert Millikan had suggested, with the latitude effect being due to Earth's magnetic field.

During World War II, Compton was a key figure in the Manhattan Project that developed the first nuclear weapons. His reports were important in launching the project. In April 1941, Vannevar Bush, head of the wartime National Defense Research Committee (NDRC), created a special committee headed by Compton to report on the NDRC uranium program. Compton's report, which was submitted in May 1941, foresaw the prospects of developing radiological weapons, nuclear propulsion for ships, and nuclear weapons using uranium-235 or the recently discovered plutonium. In October he wrote another report on the practicality of an atomic bomb. For this report, he worked with Enrico Fermi on calculations of the critical mass of uranium-235, conservatively estimating it to be between 20 kilograms (44 lb) and 2 tonnes (2.0 long tons; 2.2 short tons). He also discussed the prospects for uranium enrichment with Harold Urey, spoke with Eugene Wigner about how plutonium might be produced in a nuclear reactor, and with Robert Serber about how the plutonium produced in a reactor might be separated from uranium. His report, submitted in November, stated that a bomb was feasible, although he was more conservative about its destructive power than Mark Oliphant and his British colleagues.

The final draft of Compton's November report made no mention of using plutonium, but after discussing the latest research with Ernest Lawrence, Compton became convinced that a plutonium bomb was also feasible. In December, Compton was placed in charge of the plutonium project. He hoped to achieve a controlled chain reaction by January 1943, and to have a bomb by January 1945. In 1942, he had the research groups working on plutonium and nuclear reactor design at Columbia University, Princeton University and the University of California, Berkeley, concentrated together as the Metallurgical Laboratory in

Chicago. Its objectives were to produce reactors to convert uranium to plutonium, to find ways to chemically separate the plutonium from the uranium, and to design and build an atomic bomb.

In June 1942, the United States Army Corps of Engineers assumed control of the nuclear weapons program and Compton's Metallurgical Laboratory became part of the Manhattan Project. That month, Compton gave Robert Oppenheimer responsibility for bomb design. It fell to Compton to decide which of the different types of reactor designs that the Metallurgical Laboratory scientists had devised should be pursued, even though a successful reactor had not yet been built. When labor disputes delayed construction of the Metallurgical Laboratory's new home in the Red Gate Woods, Compton decided to build Chicago

Pile-1, the first nuclear reactor, under the stands at Stagg Field. Under Fermi's direction, it went critical on December 2, 1942. The Metallurgical Laboratory was also responsible for the design and operation of the X-10 Graphite Reactor at Oak Ridge, Tennessee.

A major crisis for the plutonium program occurred in July 1943, when Emilio Segrè's group confirmed that plutonium created in the X-10 Graphite Reactor at Oak Ridge contained high levels of plutonium-240. Its spontaneous fission ruled out the use of plutonium in a gun-type nuclear weapon. Oppenheimer's Los Alamos Laboratory met the challenge by designing and building an implosion-type nuclear weapon.

Compton was at the Hanford site in September 1944 to watch the first reactor being brought online. The first batch of uranium slugs was fed into Reactor B at Hanford in November 1944, and shipments of plutonium to Los Alamos began in February 1945. Throughout the war, Compton would remain a prominent scientific adviser and administrator. In 1945, he served, along with Lawrence, Oppenheimer, and Fermi, on the Scientific Panel that recommended military use of the atomic bomb against Japan. He was awarded the Medal for Merit for his services to the Manhattan Project.

After the war ended, Compton resigned his chair as Charles H. Swift Distinguished Service Professor of Physics at the University of Chicago and returned to Washington University in St. Louis, where he was inaugurated as the university's ninth chancellor in 1946. Compton retired as chancellor in 1954, but remained on the faculty as Distinguished Service Professor of Natural Philosophy until his retirement from the full-time faculty in 1961. In retirement he wrote *Atomic Quest*, a personal account of his role in the Manhattan Project, which was published in 1956.

Before his death, he was professor-at-large at the University of California, Berkeley for spring 1962. Compton died in Berkeley, California, from a cerebral hemorrhage on March 15, 1962.

Baade, W. (1938). The Absolute Photographic Magnitude of Supernovae.

Astrophys. J., 88, 285-304; https://articles.adsabs.harvard.edu/pdf/1938ApJ....88..285B.*

> * Contributions from the Mount Wilson Observatory, Carnegie Institution of Washington, No. 600.

Abstract

A compilation of the *photometric data* for the 18 *supernovae* known at the end of 1937 is given. Former estimates have been replaced by *photometric magnitudes* after a redetermination of the magnitudes of comparison stars on the international system. The *mean absolute photographic magnitude* of the *supernovae*, derived from this material, is $M^-_{max} = -14.3 \pm 0.42$ (m. e.) with a dispersion $M_{max} \approx 1.1$ mag. This result, together with the spectroscopic evidence, fully confirms the view that two classes of *novae, common novae* and *supernovae*, exist. Attention is drawn to the curious fact that 72 per cent of the known *supernovae* appeared in late-type spirals. *B Cassiopeiae* and the *Crab nebula*, which may have been galactic *supernovae*, are discussed.

When, after Ritchey's discovery of a *nova* in the *spiral nebula* NGC 6946 in 1917, *novae* were discovered in rapid succession in other extragalactic systems, a new way had been opened to settle the old question as to the constitution and the distances of these systems. The occurrence of *novae* in them afforded strong evidence for their stellar constitution. Moreover, their distances could be measured as soon as reliable values for the *luminosities* of the *novae* of our own galaxy were available. Nevertheless, the first applications of this method by H. D. Curtis and K. Lundmark were not very satisfactory, because the new data clearly revealed an unexpected feature, a large dispersion in the *luminosities* of the *extragalactic novae*.

This feature is well illustrated by the *novae* of the *Andromeda nebula*. The first observed in this near-by extragalactic system, the famous *nova* of 1885, *S Andromedae*, reached a maximum apparent visual *brightness* $m_v = 7.2$, a *brightness* nearly comparable with that of the whole *nebula*. On the other hand, the many *novae* discovered in the *Andromeda nebula* after a systematic search had begun in 1917 were all faint objects, ranging in *apparent magnitude* from 16 to 18, with a *frequency maximum* around 17.4. The unusually large range of 11 mag. in the observed *luminosities* of these *novae* was disturbing because *it indicated either a very large dispersion in the absolute magnitudes or the existence of two groups of novae differing in luminosity by a factor of the order 10,000.* A few years later Hubble's investigation of the *Andromeda nebula* left no doubt that the second alternative had to be adopted. The 85 faint *novae* observed in this *nebula* between 1917 and 1927 had all the characteristics of a well-defined group. Their *brightness* at maximum showed a range of only 3 or 4 mag., with a pronounced *frequency* maximum at $m = 17.3$. Moreover, their mean absolute magnitude, $M = -5.7$, identified them with the *galactic novae*[1].

[1] Additional proof that the faint *novae* of the *Andromeda nebula* are identical with the *novae* of our galaxy is afforded by later spectroscopic observations of M. L. Humason, who showed that they exhibit the typical *nova* spectrum (*Pub. A. S. P.*, 44, 381, 1932).

Compared with these common *novae*, of which an average of 30 appear annually in the Andromeda *nebula*, *S Andromedae*, with an *absolute brightness* of $M = -15.0$, stands out as a nova *sui generis*. But *S Andromedae*, though exceptional, is by no means a singular case. For, if we exclude from the list of all *novae* which have been found in extragalactic systems the few faint *novae* discovered in other members of the *local group* (those in *Messier 33* and the *Magellanic Clouds*), which are clearly common *novae*, the remaining examples are to be classed with *S Andromedae* because they reached apparent magnitudes comparable with the integrated *apparent magnitudes* of the systems in which they appeared. From the well-established fact that the *integrated luminosities* of the *extragalactic* systems are distributed around the mean value M = – 14.2, with a rather small dispersion, it follows that the remaining *novae* reach a *maximum luminosity* of the order of M = – 14. The large difference in the absolute magnitudes of these two groups strongly suggests that we are dealing with two different classes of *novae*.

Data of a quite different nature support this view. Spectroscopic investigation of a number of recent *novae* of the *S Andromedae* type has shown that these *supernovae* exhibit a most peculiar spectrum, different from that of any other known celestial object[2].

[2] For a discussion of the spectra of *supernovae* see Minkowski, R. *Mt. W. Contr. No. 602*, soon to appear.

It can be best described as consisting of wide and partly overlapping emission bands, which differ in intensity, shape, and width. This spectrum is faithfully reproduced in all six *supernovae* for which spectrograms have been obtained so far. Since no single one of the characteristic bands has been identified as to its origin, *the spectrum is a mystery at present*. And yet, as a spectroscopic definition of a *supernova*, it serves admirably, for its characteristic features are easily recognized even on spectrograms of very low dispersion. In practice, therefore, we can always decide whether a *nova* is of the common type or a *supernova* as soon as its spectrum is available. The totally different character of the spectra of common *novae* and *supernovae* puts it beyond doubt that we are dealing with two different classes of objects which, notwithstanding the similarities in their *light-curves*, probably differ radically in their underlying processes.

Preliminary estimates of the *absolute magnitudes* of Supernovae at maximum have been published by K. Lundmark[3] and by W. Baade and F. Zwicky[4].

[3] (1935). *Lund Medd.*, 2, 74.

[4] (1934). *Mt. W. Comm.,* No. 114; (1934). *Proc. Nat. Acad.*, 20, 254; (1934). *Ann Rept. Mt. W. Obs.*, p. 151.

A new determination of this constant, based on a revision of the published observational data, is presented in the following pages. In view of the meager and incomplete material now available, the result must be considered as a first approximation. But probably a number of years must elapse before better data will be at our disposal as a result of the systematic search for *supernovae* which is now under way.

The List of Supernovae.

...

Burrows. A. S.* (February, 2015). Baade and Zwicky: "Super-novae," neutron stars, and cosmic rays.

Proc. Natl. Acad. Sci., 112, 5, 1241-2; https://www.pnas.org/doi/epdf/ 10.1073/pnas.1422666112.

* Department of Astrophysical Sciences, Princeton University, Princeton, NJ.

In 1934, two astronomers in two of the most prescient papers in the astronomical literature coined the term *"supernova"*, hypothesized the existence of *neutron stars*, and knit them together with the origin of *cosmic-rays* to inaugurate one of the most surprising syntheses in the annals of science.

From the vantage point of 80 years, the centrality of *supernova* explosions in astronomical thought would seem obvious. *Supernovae* are the source of many of the elements of nature, and their blasts roil the interstellar medium in ways that inaugurate and affect star formation and structurally alter the visible component of galaxies at birth. They are the origin of most *cosmic-rays*, and these energetic rays have pronounced effects in the galaxy, even providing an appreciable fraction of the human radiation doses at the surface of the Earth and in jet flight. Prodigiously bright *supernovae* can be seen across the Universe and have been used to great effect to take its measure, and a majority of them give birth to impressively dense *neutron stars* and *black holes*. Indeed, the radio and X-ray pulsars of popular discourse, novels, and movies are rapidly spinning neutron stars injected into the galaxy upon the eruption of a supernova (Fig. 1).

However, it was only with the two startlingly prescient PNAS papers by Baade and Zwicky[1,2] in 1934 that the special character of *"super-novae"* (a term used for the first time in these papers) was highlighted, their connection with *cosmic rays* postulated, and the possibility of compact *neutron stars* hypothesized.

[1] Baade, W., & Zwicky, F. (1934). On super-novae. *Proc. Natl. Acad. Sci.*, 20, 5, 254–9.

[2] Baade, W., & Zwicky, F. (1934). Cosmic rays from super-novae. *Proc. Natl. Acad. Sci.*, 20, 5, 259–63.

(In the winter of 1933, Baade and Zwicky presented a preliminary version of these ideas at the American Physical Society Meeting at Stanford University.) To be sure, as early as 1921, in the famous Shapley–Curtis debate on the scale of the universe, Heber Curtis had stated that a division of *novae* into two magnitude classes "is not impossible".

[A *nova* (pl. *novae* or *novas*) is a transient astronomical event that causes the sudden appearance of a bright, apparently "new" star (hence the name "nova", Latin for "new") that slowly fades over weeks or months. All observed *novae* involve *white dwarfs* in close binary systems, but causes of the dramatic appearance of a *nova* vary, depending on the circumstances of the two progenitor stars. The main sub-classes of *novae* are *classical novae*, *recurrent novae* (RNe), and *dwarf novae*. They are all considered to be cataclysmic variable stars.

Classical *nova* eruptions are the most common type. This type is usually created in a close binary star system consisting of a *white dwarf* and either a *main sequence*, *subgiant*, or *red giant star*. If the orbital period of the system is a few days or less, the *white dwarf* is close enough to its companion star to draw accreted matter onto its surface, creating a dense but shallow atmosphere. This atmosphere, mostly consisting of hydrogen, is heated by the hot *white dwarf* and eventually reaches a critical temperature, causing ignition of rapid runaway fusion. The sudden increase in energy expels the atmosphere into interstellar space, creating the envelope seen as visible light during the *nova* event. In past centuries such an event was thought to be a new star. A few *novae* produce short-lived *nova remnants*, lasting for perhaps several centuries.

Under certain conditions, mass accretion can eventually trigger runaway fusion that destroys the *white dwarf* rather than merely expelling its atmosphere. In this case, the event is usually classified as a *Type Ia supernova*.

A *white dwarf* is a stellar core remnant composed mostly of electron-degenerate matter. A *white dwarf* is very dense: its mass is comparable to the Sun's, while its volume is comparable to Earth's. A *white dwarf*'s low *luminosity* comes from the emission of residual thermal energy; no fusion takes place in a white dwarf. The nearest known white dwarf is Sirius B, at 8.6 light years, the smaller component of the Sirius binary star.

White dwarfs are thought to be the final evolutionary state of stars whose mass is not high enough to become a *neutron star* or *black hole*. This includes over 97% of the stars in the Milky Way. After the hydrogen-fusing period of a *main-sequence star* of low or medium mass ends, such a star will expand to a *red giant* during which it fuses helium to carbon and oxygen in its core by the triple-alpha process. If a *red giant* has insufficient mass to generate the core temperatures required to fuse carbon (around 1 billion K), an inert mass of carbon and oxygen will build up at its center. After such a star sheds its outer layers and forms a planetary nebula, it will leave behind a core, which is the remnant *white dwarf*. Usually, *white dwarfs* are composed of carbon and oxygen.]

However, before the Baade and Zwicky papers, astronomers had not developed the idea that *supernovae*, such as S Andromedae and the bright event studied by Tycho Brahe in 1572, must be distinguished from the more common *novae*. Moreover, before these papers, the concept of a dense "*neutron star*" the size of a city but with the mass of a star like the Sun, did not exist. In their own words (italics in original)[2]: "With all reserve we advance the view that a *super-nova* represents the transition of an ordinary star into a *neutron star*, consisting mainly of *neutrons*.

Such a star may possess a very small radius and an extremely high density." In addition, the energetic class of explosions identified in the first paper (1) as "*supernovae*" naturally suggested to the authors in their second paper (2) that they could be the seat of production of the energetic particles discovered by Hess in 1911[4].

[4] Hess, V. F. (1912) Über Beobachtungen der durchdringenden Strahlung bei sieben Freiballonfahrten. *Phys. Zeit.*, 13, 1084–91.

Baade and Zwicky state (2): "We therefore feel justified in advancing tentatively the hypothesis that *cosmic rays are produced in the supernova process*" (italics in original). Eighty years later, this remains the view of astrophysicists.

...

Fig. 1. A picture of the inner regions of the famous Crab Nebula captures emergent jets and the "Napoleon Hat" structure of surrounding plasma. The radio/optical/X-ray pulsar, a neutron star rotating at ~30 Hz, is buried in the center. The Crab was produced in a supernova explosion in A.D. 1054. Image courtesy of ESA/NASA. [See front cover.]

The concept of a *supernova* was rapidly accepted, and in the following years many examples were found. After all, the out-sized blast waves that are the "*supernova* remnants" in our galaxy (Fig. 2), and the explosive transients seen in other galaxies ("island universes") that astronomers had recently demonstrated were outside our galaxy and distant, had therefore to be extraordinarily energetic. However, the concept of a *neutron star* was initially met with skepticism, despite the theoretical calculations of Oppenheimer and Volkoff[9], and it was not until the discovery of *radio pulsars* in 1967[10] more than 30y later—and their interpretation as spinning *neutron stars* the next year (11)—that the concept of a *neutron star* was accepted and mainstreamed.

[9] Oppenheimer, J. R., & Volkoff, G. M. (1939). On massive neutron cores. *Phys. Rev.*, 55, 4, 374–81.

[10] Hewish, A., Bell, S. J., Pilkington, J. D. H., Scott, P. F., & Collins, R. (1968). Observation of a rapidly pulsating radio source. *Nature*, 217, 5130, 709–13.

Today, we know of many thousands of *radio pulsars* and *neutron star* systems, and their study engages many in the astronomical community.

As might have been anticipated, most of the quantitative results presented in the Baade and Zwicky papers from 1934[1,2] have not survived. However, the authors were motivated to posit a *neutron star* by the extraordinary energy they concluded was required to explain their *supernovae*, and to produce energetic *cosmic rays* simultaneously, impulsively, and copiously. A *neutron* star would be very dense and, in the words of Baade and Zwicky, the "gravitational packing energy" would be very high[2]. The authors had eliminated nuclear energy as too small to power a *supernova*, and believed they needed a nontrivial fraction of the *rest-mass energy* of the star. (Note also that the year 1934 was before we fully understood the nuclear processes that power stars.) This fraction Baade and Zwicky could obtain from the *gravitational binding energy* of a compact object with nuclear or greater densities. The *neutron* had just been discovered in 1932[12] and was known to be neutral, and Baade and Zwicky imagined that oppositely charged *protons* and *electrons* could be crushed together to produce their beast.

[12] Chadwick, J. (1932). Possible existence of a neutron. *Nature*. 129, 312.

The modern view[13] is not extravagantly different, although one now quotes Baade and Zwicky for profound insight, not technical accuracy.

[13] Burrows, A. (2013) Perspectives on core-collapse supernova theory. *Rev. Mod. Phys.*, 85, 1, 245–61.

Importantly, one type of *supernova*, the *Type Ia*, is indeed powered by *nuclear energy*. In fact, and ironically, all of the supernovae observed by Baade and Zwicky in the 1930s were of this type, not of the majority type currently thought to be powered ultimately by *gravitation*.

Many believe that Lev Landau predicted the existence and characteristics of *neutron stars* soon after the discovery of the *neutron*[14].

[14] Landau, L. D. (1932). On the theory of stars. *Phys. Z. Sowjetunion*, 1, 285–88.

However, as Yakovlev *et al.*[15] have clearly shown, Laudau was thinking about a macroscopic nucleus and nowhere in that paper was the *neutron* mentioned.

[15] Yakovlev, D. G., Haensel, P., Baym, & G., Pethick, C. (2013). Lev Landau and the concept of neutron stars. *Physics Uspekhi*, 56, 289–95.

Landau's paper[14] was in fact written before the discovery of the *neutron*, and incorporated the misunderstanding that *quantum mechanics* for nuclear processes required the violation of energy conservation. Hence, the appearance of Landau's paper in 1932 was a coincidence. However, Landau did address what is now known as the *"Chandrasekhar mass"* for *white dwarfs*, and his concept of a *compact star* was a creative departure.

More than 250,000 papers have been written since, with either the words *"supernova"* or *"neutron star"* in their title or abstract (according to NASA's Astro-physics Data System, adsabs.harvard.edu/abstract_service.html). Four Nobel Prizes in Physics have been awarded for work involving *supernovae* and *neutron stars* in some way. As of 2014, more than 6,500 *supernovae* have been discovered. The theory of *cosmic-ray* acceleration in *supernova* remnants is now a well-developed topic in modern astrophysics. However, the leap of imagination shown by Baade and Zwicky in 1934 in postulating the existence of two new classes of astronomical objects, and in connecting three now central astronomical fields into one whole, still leaves one breathless. Even decades later, such a reaction continues to be a fitting tribute to these landmark PNAS papers[1,2].

Tonry, J.* & Schneider, D. P. [†] (September, 1988). A New Technique for Measuring Extragalactic Distances[§]

Astron. J., 96, 3, 807; https://articles.adsabs.harvard.edu/pdf/
1988AJ.....96..807T.

* Physics Department, Massachusetts Institute of Technology, Cambridge, Massachusetts.

[†] Institute for Advanced Study, Princeton, New Jersey.

[§] Observations taken at the Palomar Observatory, California Institute of Technology, and at the McGraw Hill Observatory, operated jointly by the University of Michigan, Dartmouth College, and the Massachusetts Institute of Technology.

This work was partially funded by NASA grants nos. NAS5-29225 and NSG7643, and an Exxon fellowship.

Received April 26, 1988.

Abstract

We describe a relatively direct technique of determining *extragalactic distances*. The method relies on measuring the *luminosity fluctuations* that arise from the *counting statistics* of the *stars* contributing the *flux* in each pixel of a high-signal-to-noise CCD (*Charge Coupled Device*) image of a *galaxy*. *The amplitude of these fluctuations is inversely proportional to the distance of the galaxy.* This approach bypasses most of the successive stages of calibration required in the traditional *extragalactic distance ladder*; the only serious drawback to this method is that it requires an accurate knowledge of the bright end ($M_v < 3$) of the *luminosity function*. Potentially, this method can produce accurate distances of *elliptical galaxies* and *spiral bulges* at distances out to about 20 Mpc. In this paper, we explain how to calculate the value of the fluctuations, taking into account various sources of contamination and the effects of finite spatial resolution, and we demonstrate, via simulations and CCD images of M32 and N3379, the feasibility and limitations of this technique.

I. INTRODUCTION

A basic problem in astrophysics is the measurement of the *distance* to celestial objects. Truly reliable distances can be derived from radar timing in the solar system and parallax measurements of nearby *stars*. For more distant sources, we must rely on indirect techniques such as the moving cluster method, statistical parallaxes, or main-sequence fitting (see, for example, Mihalas and Binney 1981). The traditional approach for *extragalactic* measurements is to construct a distance ladder of ever more luminous and rare beacons that can be seen to enormous distances, stand out from the surrounding background, and which, we trust, have constant or calibratable *luminosities*. As we bootstrap ourselves up this distance ladder, however, each successive rung is progressively more uncertain; the estimators become more exotic and less well understood, and the propagation of systematic errors produces large cumulative uncertainties after just a few steps.

A different approach is suggested by Fig. 1 [Plate 33]. This high-signal-to-noise picture of the nearby elliptical *galaxy* M32 reveals a striking mottling that is reminiscent of Baade's (1944) classic resolution of scattered, very bright *stars* in M31. It is easy to show, however, that, unlike Baade, we are seeing tens of *stars* per square arcsecond that lie on the ordinary giant branch rather than individual, extremely *luminous supergiants*. We were therefore led to consider using the amplitude of the mottling as a distance estimator, since it will decrease inversely with distance. The measurement of these "*luminosity fluctuations*" has the potential to be a very powerful approach, for it bypasses many of the intermediate steps used by previous methods; the only calibration needed is a knowledge of the *stellar luminosity function*. There are a number of additional advantages to this method, such as high accuracy, ease of observation, and well-understood objects contributing to the *luminosity*.

The primary disadvantages are that the *stellar luminosity function* must be quite accurately known and that high-signal-to-noise data are required; the latter item demands that the contributors to the noise and fluctuations in the data be well understood. The finite spatial resolution introduced by instrumental and atmospheric blurring, while setting the ultimate limits for the technique, is actually an asset to the method in some respects.

The following section investigates the sources and amplitudes of *luminosity fluctuations* and describes the steps required for data processing. Section III illustrates the expected behavior of *galaxies* at different distances, and shows the results of this technique when applied to simulated data. Observations of M32 and N3379 are presented in Sec. IV. In Sec. V we discuss the results and implications, and suggest directions for future research.

II. MEASUREMENT OF LUMINOSITY FLUCTUATIONS

The idea behind *luminosity fluctuations* is simple. Imagine a *galaxy* made of only one type of *star*, ignore seeing for the moment, and suppose that at some region in the *galaxy* the average projected density of *stars* is one hundred per pixel. We do not resolve individual *stars*, but as we look at adjacent pixels, we will see fluctuations with rms variations equal to 10% of the mean signal. At different locations in the *galaxy*, we will see rms fluctuations that vary as the square root of the local mean *galaxy brightness*; the proportionality constant between fluctuation rms and square root of the mean surface brightness will be directly related to the number of stars present. If the *intrinsic luminosity* of the stars is known, it is an easy matter to calculate the distance of the *galaxy* from the measured fluctuations and the observed *surface brightness*. Now consider an identical *galaxy* that is twice as distant. If we look at the first location where there used to be one hundred per pixel, there are now 400 *stars* per pixel because the pixel subtends 4 times the metric area (the flux from each *star*, of course, is decreased by a factor of 4, so the *surface brightness* is constant). In this *galaxy*, the fluctuations are 5% at this flux level. Different regions in the *galaxy* will have a linear relation between the rms fluctuations and square root of *galaxy flux*, but the proportionality constant will be half as large as for the first *galaxy*; this constant is inversely proportional to distance.

(a) *Sources of Fluctuations*

In reality, there are many sources of pixel-to-pixel fluctuations in a CCD image of a *galaxy*. Some, like *globular clusters, dwarf companions,* or the *fluctuations* that we seek to measure and employ, are essentially intrinsic to the *galaxy* in question. Others involve foreground *stars* or background *galaxies*, and yet others arise from various kinds of instrumental noise. The total variance (or mean-square value) at any point in the image will be the sum of the variances of each component, under the reasonable assumption of uncorrelated phases.

We will shortly catalog and characterize these wanted and unwanted variances, but already there is one vital point to keep in mind: the *spatial power spectrum of fluctuations* is the key to this method. First, it will allow us to determine the amplitude of fluctuations that have been degraded by atmospheric blurring. Second, it will give us the means to distinguish many instrumental sources of noise (which have a white power spectrum) from fluctuations intrinsic to the galaxy (which will have the point-spread function (psf) impressed upon them). If we restrict ourselves to early-type *galaxies* and zealously excise point sources, we can avoid fluctuations from clumping of *stars*, from patchy obscuration, or from objects such as *globular clusters, H_{II} regions, planetary nebulae, companion galaxies*, etc. The feasibility of this will be demonstrated below.

...

V. DISCUSSION

...

We believe that this is the first quantitative measurement of *luminosity fluctuations*, but the relationship between the resolution of *stars*, the resulting fluctuations, and the distance of the galaxy has occurred to many people before us. We are aware that William Baum has thought a great deal about this topic, particularly insofar as the resolution of *Hubble Space Telescope* (HST) will improve matters. This idea also was related to one of us (D.P.S.) by Peter Young, almost a decade ago. In many respects, this technique has only been waiting for the superb qualities of modern CCDs to come to fruition. ...

The expanding universe and the Big Bang.

The *Big Bang* is a physical theory that describes how the universe expanded from an initial state of high density and temperature.

In 1912, Vesto Slipher measured what was believed to be the first Doppler shift of a "*spiral nebula*" (spiral nebula is the obsolete term for spiral galaxies), and soon "discovered" that almost all such nebulae were receding from Earth. He did not grasp the cosmological implications of this fact, and indeed at the time it was highly controversial whether or not these nebulae were "island universes" outside our Milky Way.

Ten years later, in 1922, Alexander Friedmann, a Russian cosmologist and mathematician, derived the Friedmann equations from the field equations of *Einstein's theory of general relativity*, showing that the universe might be expanding in contrast to the static universe model advocated by Albert Einstein at that time.

Independently deriving Friedmann's *relativistic* equations in 1927, Georges Lemaître, a Belgian physicist and Roman Catholic priest, proposed that the recession of the nebulae was due to the expansion of the universe. He inferred the relation that Hubble would later observe, given the *cosmological principle*.

> [The *cosmological principle* states that the universe is homogeneous and isotropic on a large scale. This means that the distribution of matter and energy is uniform, and the universe looks the same in all directions. It asserts that there are no preferred locations or directions in space.]

In 1924, American astronomer Edwin Hubble's measurement of the great distance to the nearest *spiral nebulae* showed that these systems were indeed other galaxies. Starting that same year, Hubble painstakingly developed a series of distance indicators, the forerunner of the *cosmic distance ladder*, using the 100-inch (2.5 m) Hooker telescope at Mount Wilson Observatory. This allowed him to estimate *distances* to galaxies whose *redshifts* had already been measured, mostly by Slipher. In 1929, Hubble discovered a correlation between *distance* and *redshift*, described as *radial velocity*—now known as Hubble's law.

In 1931, Lemaître went further and suggested that the evident expansion of the universe, if projected back in time, meant that the further in the past the smaller the universe was, until at some finite time in the past all the mass of the universe was concentrated into a single point, a "primeval atom" where and when the fabric of time and space came into existence.

In the 1920s and 1930s, almost every major cosmologist preferred an eternal steady-state universe. During the 1930s, other ideas were proposed as non-standard cosmologies to explain Hubble's observations, including the Milne model, the oscillatory universe (originally suggested by Friedmann, but advocated by Albert Einstein and Richard C. Tolman), and Fritz Zwicky's *"tired-light" hypothesis*.

After World War II, two distinct possibilities emerged. One was Fred Hoyle's *steady-state model*, whereby new matter would be created as the universe *seemed to expand*. In this model the universe is roughly the same at any point in time. The other was Lemaître's *Big Bang theory*, advocated and developed by George Gamow, who introduced *Big Bang nucleosynthesis* (BBN).

[*Big Bang nucleosynthesis* (BBN) is the production of nuclei other than those of the lightest isotope of hydrogen (^{1}H, having a single proton as a nucleus) during the early phases of the universe. This type of *nucleosynthesis* was believed to have occurred from 10 seconds to 20 minutes after the *Big Bang*. It was thought to be responsible for the formation of most of the universe's helium (as isotope ^{4}He), along with small fractions of the hydrogen isotope deuterium (^{2}H or D), the helium isotope helium ^{3}He, and a very small fraction of the lithium isotope lithium ^{7}Li. Elements heavier than lithium were thought to have been created later in the life of the Universe by *stellar nucleosynthesis*, through the formation, evolution and death of stars.]

Ironically, it was Hoyle who coined the phrase that came to be applied to Lemaître's theory, referring to it as "this big bang idea" during a BBC Radio broadcast in March 1949. For a while, support was split between these two theories. The observational evidence, most notably from radio source counts, began to favor *Big Bang* over *steady state*, but in the mid-1990s, observations of certain globular clusters appeared to indicate that they were about 15 billion years old, which conflicted with most then-current estimates of the age of the universe.

Various cosmological models of the Big Bang attempted to explain the evolution of the observable universe from the earliest known periods through its subsequent large-scale form. The models depended on two major assumptions: the *universality of physical laws* and the *cosmological principle*. The *universality of physical laws* is one of the underlying principles of the *theory of relativity*. The *cosmological principle* states that on large scales the universe is homogeneous and isotropic—appearing the same in all directions regardless of location.

They offered an explanation for a broad range of observed phenomena, including the abundance of light elements, the *cosmic microwave background* (CMB) radiation, and large-scale structure. These models were compatible with the Hubble–Lemaître law—the observation that *the farther away a galaxy is, the faster it appeared to be moving away from Earth*. Extrapolating this cosmic expansion backward in time using the known laws of physics, the models describe an increasingly concentrated cosmos preceded by a singularity in which space and time lose meaning (the "*Big Bang singularity*"). Detailed measurements of the expansion rate of the universe placed the *Big Bang singularity* at an estimated 13.8 billion years ago, which was considered the *age of the universe*.

However, physics lacked a widely accepted theory of *quantum gravity* that can model the earliest conditions of the *Big Bang*, and there are other aspects of the observed universe that were not adequately explained by the *Big Bang models*. After its initial expansion, the universe cooled sufficiently to allow the formation of subatomic particles, and later atoms. The unequal abundances of *matter* and *antimatter* that allowed this to occur is an unexplained effect known as *baryon asymmetry*. These primordial elements—mostly hydrogen, with some helium and lithium— were believed to later coalesce through gravity, forming early stars and galaxies. Astronomers observed the gravitational effects of an *unknown dark matter* surrounding galaxies. Most of the *gravitational potential* in the universe seems to be in this form, and the *Big Bang models* and various observations indicated that this excess *gravitational potential* was not created by *baryonic matter*, such as normal atoms. Measurements of the *redshifts* of *supernovae* indicated that the expansion of the universe was accelerating, an observation attributed to an *unexplained* phenomenon known as *dark energy*.

However, according to the analysis in this volume, it now appears that the *redshifts*, on which the *expansion of the universe* and the "*Big Bang*" were based, were not caused by the Doppler effect on light from receding galaxies but rather were the result of the interaction of the light from galaxies with electrons and other matter as it travelled through *intergalactic space*. Zwicky's *"tired-light" theory* was correct.

> Shamir, L. (August, 2024). *An Empirical Consistent Redshift Bias: A Possible Direct Observation of Zwicky's TL Theory* (see below): "The early Universe as imaged by JWST is different from the early Universe predicted by the standard model. Among other explanations, that tension was proposed to have a link to Fritz Zwicky's *"tired-light" theory* (TL). According to TL, *photons* lose their energy along their traveling path through the Universe. This can lead to differences in the *redshift* as observed from Earth, and, therefore, galaxies that are more distant from Earth can have *redshift* that differs from galaxies that are closer to Earth. If this theory is correct, it would explain the early mature galaxies observed by JWST, as their true age would not the same

age that their *redshift* indicates. *In its extreme form, Zwicky's "tired-light" theory (TL) can argue that the Big Bang is merely an artifact created by TL, and that the Universe is in fact in a steady state."*

Perlmutter, S.*, Aldering, G., Goldhaber, G.*, Knop, R.A., Nugent, P., *et al.* (December, 1998). Measurements of Omega and Lambda from 42 High-Redshift Supernovae.

Astrophys. J., 517, 565-86; https://arxiv.org/pdf/astro-ph/9812133.

* Center for Particle Astrophysics, U.C. Berkeley, California.

This paper reports on the *Supernova Cosmology Project*, which was started in 1988 to determine the cosmological parameters of the universe using the *magnitude-redshift* relation of *Type Ia supernovae*. All *supernova* peak magnitudes are standardized using a *SN Ia light-curve width-luminosity relation*. It determines that the data are *strongly inconsistent* with a $\Lambda = 0$ *flat cosmology*, the simplest inflationary universe model in which the universe continues to expand at a *constant rate*, for which the best-fit age of the universe relative to the Hubble time was $t_0^{flat} = 14.9_{-1.1}^{+1.4}$ (0.63/h) Gyr. The data indicate that the *cosmological constant* is *non-zero* and *positive*, indicating an *accelerating universe*. This relation is based on a *relativistic cosmological model*, so is inconsistent with New Physics.

Abstract

We report measurements of the *mass density*, Ω_M, and *cosmological-constant energy density*, Ω_Λ, of the universe based on the analysis of 42 *Type Ia supernovae* discovered by the *Supernova Cosmology Project*. The *magnitude-redshift* data for these *supernovae*, at *redshifts* between 0.18 and 0.83, are fit jointly with a set of *supernovae* from the Calán/Tololo Supernova Survey, at *redshifts* below 0.1, to yield values for the cosmological parameters. All *supernova* peak magnitudes are standardized using a *SN Ia light-curve width-luminosity relation*. The measurement yields a joint probability distribution of the cosmological parameters that is approximated by the relation $0.8\,\Omega_M - 0.6\,\Omega_\Lambda \approx -0.2 \pm 0.1$ in the region of interest ($\Omega_M < 1.5$). For a *flat* ($\Omega_M + \Omega_\Lambda = 1$)

cosmology we find $\Omega^{flat}_M = 0.28^{+0.09}_{-0.08}$ (1σ statistical) $^{+0.05}_{-0.04}$ (identified systematics).

The data are *strongly inconsistent with a* $\Lambda = 0$ *flat cosmology, the simplest inflationary universe model.*

[The *cosmological constant* (lambda: Λ), alternatively called *Einstein's cosmological constant*, is a coefficient that Albert Einstein initially added to his *field equations* of *general relativity*. He later removed it; however, much later it was revived to express the *energy density* of space, or *vacuum energy*, that arises in *quantum mechanics*. It is closely associated with the concept of *dark energy*.

Einstein introduced the constant in 1917 *to counterbalance the effect of gravity and achieve a static universe,* which was then assumed. *Einstein's cosmological constant* was abandoned after Hubble's measurements suggested that the universe was expanding. From the 1930s until the late 1990s, most physicists agreed with Einstein's choice of setting *the cosmological constant* to zero. That changed with the claim in 1998 that the *expansion of the universe* is accelerating, implying that the *cosmological constant* may have a positive value.

Since the 1990s, studies have shown that, assuming the *cosmological principle*, around 68% of the *mass–energy density* of the universe can be attributed to *dark energy*. The *cosmological constant* Λ is the simplest possible explanation for *dark energy*, and is used in the *standard model* of cosmology known as the *ΛCDM model*.

The *cosmological principle* is the notion that the spatial distribution of matter in the universe is uniformly isotropic and homogeneous when viewed on a large enough scale, since the forces are expected to act equally throughout the universe on a large scale, and should, therefore, produce no observable inequalities in the large-scale structuring over the course of evolution of the matter field that was initially laid down by the *Big Bang*.]

An *open*, $\Lambda = 0$, *cosmology* also does not fit the data well: the data indicate that the *cosmological constant* is *non-zero* and *positive*, with a confidence of $P(\Lambda > 0) = 99\%$, including the identified systematic uncertainties.

[An *open cosmology* with the *cosmological constant* $\Lambda = 0$ describes a universe that continues to expand at a constant rate.]

The best-fit *age of the universe* relative to the *Hubble time* is $t_0^{\text{flat}} = 14.9_{-1.1}^{+1.4}$ (0.63/h) Gyr for a *flat cosmology*.

[The motion of astronomical objects is described by the equation $v = H_0 D$, with H_0 the constant of proportionality—the *Hubble constant*—between the "*proper distance*" D to a galaxy (which can change over time, unlike the comoving distance) and its *speed of separation* v, i.e. the derivative of *proper distance* with respect to the *cosmic time coordinate*.

The *Hubble constant* is most frequently quoted in km/s/Mpc, which gives the speed of a galaxy 1 megaparsec (3.09×10^{19} km) away as 70 km/s. Simplifying the units of the generalized form reveals that H_0 specifies a frequency (SI unit: s^{-1}), leading the reciprocal of H_0 to be known as the *Hubble time* (14.4 billion years).

Flatness is a property of a *space without curvature*. Such a space is called a "*flat space*" or *Euclidean space*. Whether the universe is "*flat*" could determine its ultimate fate; whether it will expand forever, or ultimately collapse back into itself. The geometry of *spacetime* has been measured by the Wilkinson Microwave Anisotropy Probe (WMAP) to be nearly *flat*. According to the WMAP 5-year results and analysis, "*WMAP determined that the*

111

universe is flat, from which it follows that the *mean energy density* in the universe is equal to the *critical density* (within a 1% margin of error). This is equivalent to a *mass density* of 9.9×10^{-30} g/cm^3, which is equivalent to only 5.9 protons per cubic meter." The WMAP data are consistent with a *flat geometry*, with the *mass density*, $\Omega = 1.02 +/- 0.02$.]

The size of our sample allows us to perform a variety of statistical tests to check for possible systematic errors and biases. We find no significant differences in either the host *reddening* distribution or Malmquist bias between the *low-redshift* Calán/Tololo sample and our *high-redshift* sample. Excluding those few *supernovae* which are outliers in color excess or fit residual does not significantly change the results. The conclusions are also robust whether or not a *width-luminosity relation* is used to standardize the *supernova* peak magnitudes. We discuss, and constrain where possible, hypothetical alternatives to a *cosmological constant*.

---

1. INTRODUCTION

Since the earliest studies of *supernovae*, it has been suggested that *these luminous events might be used as standard candles for cosmological measurements*[1].

> [1] Baade, W. (1938) The absolute photographic magnitude of supernovae. *Astrophys. J.*, 88, 285–304; https://articles.adsabs.harvard.edu/pdf/1938CMWCI.600....1B.

At closer distances they could be used to measure the *Hubble constant*, if an absolute *distance* scale or *magnitude* scale could be established, while at higher *redshifts* they could determine the *deceleration parameter* (Tammann 1979; Colgate 1979).

[The *deceleration parameter* in cosmology is a dimensionless measure of the *cosmic acceleration of the expansion of space* in a *Friedmann–Lemaître–Robertson–Walker universe*.

The *Friedmann–Lemaître–Robertson–Walker metric* is a metric based on an exact solution of the *Einstein field equations of general relativity*.]

The *Hubble constant* measurement became a realistic possibility in the 1980's, when the more homogeneous subclass of *Type Ia supernovae* (SNeIa) was identified (see Branch 1998). Attempts to measure the *deceleration parameter*, however, were stymied for lack of *high-redshift supernovae*. Even after an impressive multi-year effort by Nørgaard-Nielsen *et al.* (1989), *it was only possible to follow one SN Ia, at* $z = 0.31$, *discovered 18 days past its peak brightness*.

The *Supernova Cosmology Project* was started in 1988 to address this problem. The primary goal of the project is the determination of the cosmological parameters of the universe using the *magnitude-redshift* relation of *Type Ia supernovae*. In particular, Goobar & Perlmutter (1995) showed the possibility of separating the relative contributions of the *mass density*, Ω_M, and the *cosmological constant*, Λ, to changes in the expansion rate by studying *supernovae* at a range of *redshifts*. The *Project* developed techniques, including instrumentation, analysis, and observing strategies, that make it possible to systematically study *high-redshift supernovae* (Perlmutter *et al.* 1995a). As of March 1998, more than 75 *Type Ia supernovae* at *redshifts* $z = 0.18$–0.86 have been discovered and studied by the *Supernova Cosmology Project* (Perlmutter *et al.* 1995b, 1996, 1997a,b,c,d, 1998a).

A first presentation of analysis techniques, identification of possible sources of statistical and systematic errors, and first results based on seven of these supernovae at *redshifts* $z \sim 0.4$ were given in Perlmutter *et al.* (1997e; hereafter referred to as "P97"). These first results yielded a confidence region that was suggestive of a *flat*, $\Lambda = 0$ *universe*, but with a large range of uncertainty. Perlmutter *et al.* (1998b) added a $z = 0.83$

SN Ia to this sample, with observations from the Hubble Space Telescope and Keck 10-meter telescope, providing the first demonstration of the method of separating Ω_M and Λ contributions. This analysis offered preliminary evidence for a *low-mass-density* universe with a best-fit value of $\Omega_M = 0.2 \pm 0.4$, assuming $\Lambda = 0$. Independent work by Garnavich *et al.* (1998a), based on three supernovae at $z \sim 0.5$ and one at $z = 0.97$, also suggested a *low mass density*, with best-fit $\Omega_M = -0.1 \pm 0.5$ for $\Lambda = 0$.

Perlmutter *et al.* (1998c) presented a preliminary analysis of 33 additional *high-redshift* supernovae, which gave a confidence region indicating an *accelerating universe*, and *barely including a low-mass $\Lambda = 0$ cosmology*. Recent independent work of Riess *et al.* (1998), based on 10 *high-redshift supernovae* added to the Garnavich *et al.* set, reached the same conclusion. Here we report on the complete analysis of 42 *supernovae* from the *Supernova Cosmology Project*, including the reanalysis of our previously reported *supernovae* with improved calibration data and improved photometric and spectroscopic *SN Ia* templates.

2. BASIC DATA AND PROCEDURES

The new *supernovae* in this sample of 42 were all discovered while still brightening, using the Cerro Tololo 4-meter telescope with the 2048^2-pixel prime-focus CCD camera or the 4×2048^2-pixel Big Throughput Camera (Bernstein & Tyson 1998). The *supernovae* were followed with photometry over the peak of their *light-curves*, and approximately two-to-three months further (~ 40–60 days rest-frame) using the CTIO 4-m, WIYN 3.6-m, ESO 3.6-m, INT 2.5-m, and WHT 4.2-m telescopes. (SN 1997ap and other 1998 *supernovae* have also been followed with *Hubble Space Telescope* (HST) photometry.) The *supernova redshifts* and spectral identifications were obtained using the Keck I and II 10-m telescopes with LRIS (Oke *et al.* 1995) and the ESO 3.6 m telescope. The photometry coverage was most complete in Kron-Cousins R-band, with Kron-Cousins I-band photometry coverage ranging from two or three

114

points near peak to relatively complete coverage paralleling the R-band observations.

Almost all of the new *supernovae* were observed spectroscopically. The confidence of the Type Ia classifications based on these spectra taken together with the observed *light-curves*, ranged from "definite" (when Si II features were visible) to "likely" (when the features were consistent with *Type Ia*, and inconsistent with most other types). The lower confidence identifications were primarily due to host-galaxy contamination of the spectra. Fewer than 10% of the original sample of *supernova* candidates from which these *SNe Ia* were selected were confirmed to be non-*Type Ia*, i.e., being *active galactic nuclei* or belonging to another SN subclass; almost all of these non-*SNe Ia* could also have been identified by their *light-curves* and/or position far from the *SN Ia* Hubble line. Whenever possible, the *redshifts* were measured from the narrow host-galaxy lines, rather than the broader *supernova* lines. The *light-curves* and several spectra are shown in Perlmutter *et al.* (1997e, 1998c, 1998b); complete catalogs and detailed discussions of the photometry and spectroscopy for these *supernovae* will be presented in forthcoming papers.

The photometric reduction and the analyses of the *light-curves* followed the procedures described in P97. The *supernovae* were observed with the Kron-Cousins filter that best matched the rest-frame B and V filters at the *supernova's redshift*, and any remaining mismatch of wavelength coverage was corrected by calculating the expected photometric difference— the "cross-filter K-correction"—using template *SN Ia* spectra, as in Kim, Goobar, & Perlmutter (1996). We have now recalculated these K corrections (see Nugent *et al.* 1998), using improved template spectra, based on an extensive database of *low-redshift SN Ia* spectra recently made available from the Calán/Tololo survey (Phillips *et al.* 1998). Where available, IUE and HST spectra (Cappellaro, Turatto, & Fernley 1995; Kirshner *et al.* 1993) were also added to the *SN Ia* spectra, including those published previously (1972E, 1981B, 1986G, 1990N, 1991T, 1992A, and 1994D in: Kirshner & Oke 1975; Branch *et*

115

al. 1983; Phillips *et al.* 1987; Jeffery *et al.* 1992; Meikle *et al.* 1996; Patat *et al.* 1996). In Nugent *et al.* (1998) we show that the K-corrections can be calculated accurately for a given day on the *supernova light-curve*, and for a given *super nova light-curve* width, from the color of the *supernova* on that day. (Such a calculation of K correction based on *supernova color* will also automatically account for any modification of the K correction due to *reddening* of the *supernova*; see Nugent *et al.* 1998. In the case of insignificant *reddening* the *SN Ia* template color curves can be used.) We find that these calculations are robust to mis-estimations of the *light-curve width* or day on the *light-curve*, giving results correct to within 0.01 mag for *light-curve width* errors of ± 25% or *light-curve phase errors* of ± 5 days even at *redshifts* where filter matching is the worst. Given small additional uncertainties in the colors of *supernovae*, we take an overall systematic uncertainty of 0.02 magnitudes for the K correction.

The improved K-corrections have been recalculated for all the *supernovae* used in this paper, including those previously analyzed and published. Several of the *low-redshift supernovae* from the Calán/Tololo survey have relatively large changes (as much as 0.1 magnitudes) at times in their K-corrected *light-curves*. (These and other *low-redshift supernovae* with new K-corrections are used by several independent groups in constructing SN Ia *light-curve* templates, so the templates must be updated accordingly.) The K-corrections for several of the *high-redshift supernovae* analyzed in P97 have also changed by small amounts at the *light-curve peak* ($\Delta K(t = 0) \sim < 0.1$ mag); this primarily affects the measurement of the rest frame *light-curve width*. These K-correction changes balance out among the P97 *supernovae*, so the final results for these *supernovae* do not change significantly. (As we discuss below, however, the much larger current dataset does affect the interpretation of these results.).

As in P97, the peak magnitudes have been corrected for the *light-curve width-luminosity relation* ...

5. RESULTS AND ERROR BUDGET

From Table 3 and Figure 5(a), it is clear that the results of Fits A, B, and C are quite close to each other, so we can conclude that our measurement is robust with respect to the choice of these supernova subsets. The inclusive Fits A and B are the fits with the least subjective selection of the data. They already indicate the main cosmological results from this dataset. However, to make our results robust with respect to host-galaxy reddening, we use Fit C as our primary fit in this paper. For Fit C, we find $\Omega_M^{flat} = 0.28_{-0.08}^{+0.09}$ in a *flat universe*. Cosmologies with $\Omega_\Lambda = 0$ are a poor fit to the data, at the 99.8% confidence level. The contours of Figure 7 more fully characterize the best-fit confidence regions. (The table of this two-dimensional probability distribution is available at http://www.supernova.lbl.gov/.)

The residual plots of Figure 2(b and c) indicate that the best fit Ω_M^{flat} in a *flat universe* is consistent across the *redshift* range of the *high-redshift supernovae*.

…

… The narrow intrinsic dispersion— which does not increase at *high redshift*—provides additional evidence against an increase in *extinction* with *redshift*. Even if there is grey dust *that dims the supernovae without reddening them*, the dispersion would increase, unless the dust is distributed very uniformly.

A *flat, $\Omega_\Lambda = 0$ cosmology* is a *quite poor fit to the data*. The $(\Omega_M, \Omega_\Lambda) = (1, 0)$ line shows that 38 out of 42 *high-redshift supernovae* are fainter than predicted for this model. These *supernovae* would have to be over 0.4 magnitudes brighter than measured (or the *low-redshift supernovae* 0.4 magnitudes fainter) for this model to fit the data. …

6. CONCLUSIONS AND DISCUSSION

The confidence regions of Figure 7 and the residual plot of Figure 2(b) lead to several striking implications. First, *the data are strongly inconsistent with the* $\Lambda = 0$, *flat universe model* (indicated with a circle) that has been the theoretically favored cosmology. *If the simplest inflationary theories are correct and the universe is spatially flat, then the supernova data imply that there is a significant, positive cosmological constant.* Thus, *the universe may be flat*, or *there may be little or no cosmological constant*, but the data are not consistent with both possibilities simultaneously. This is the most unambiguous result of the current dataset.

Second, this dataset directly addresses the *age of the universe* relative to the *Hubble time*, H_0^{-1}. Figure 9 shows that the $\Omega_M - \Omega_\Lambda$ confidence regions are almost parallel to contours of constant age. For any value of the *Hubble constant* less than $H_0 = 70$ km s^{-1} Mpc^{-1}, *the implied age of the universe is greater than 13 Gyr*, allowing enough time for the oldest stars in globular clusters to evolve (Chaboyer *et al.* 1998; Gratton *et al.* 1997). Integrating over Ω_M and Ω_Λ, the best fit value of the age in *Hubble-time* units is $H_0 t_0 = 0.93_{-0.06}^{+0.06}$ or equivalently $t_0 = 14.5_{-1.0}^{+1.0}$ (0.63/h) Gyr. The age would be somewhat larger in a *flat universe*: $H_0 t_0^{\text{flat}} = 0.96_{-0.07}^{+0.09}$ or equivalently $t_0^{\text{flat}} = 14.9_{-1.1}^{+1.4}$ (0.63/h) Gyr.

Third, *even if the universe is not flat, the confidence regions suggest that the cosmological constant is a significant constituent of the energy density of the universe.* The best fit model indicates that the *energy density* in the *cosmological constant* is ~ 0.5 more than that in the form of *mass energy density.* All of the alternative fits indicate a *positive cosmological constant* with confidence levels of order 99%, even with the systematic uncertainty included in the fit or with a *clumped-matter metric.*

Given the potentially revolutionary nature of this third conclusion, it is important to reexamine the evidence carefully to find possible loopholes. None of the identified sources of statistical and systematic uncertainty described in the previous sections could account for the data in a $\Lambda = 0$ universe. *If the universe does in fact have zero cosmological constant, then some additional physical effect or "conspiracy" of statistical effects must be operative* — and must make the *high-redshift super novae* appear almost 0.15 mag (~15% in flux) fainter than the *low-redshift supernovae*. At this stage in the study of *SNe Ia*, we consider this unlikely *but not impossible*. For example, as mentioned above, some carefully constructed *smooth distribution of large-grain-size grey dust that evolves similarly for elliptical and spiral galaxies could evade our current tests*. Also, the full dataset of well-studied *SNe Ia* is still *relatively small*, particularly at *low redshifts*, and we would like to see a more extensive study of *SNe Ia* in many different host-galaxy environments before we consider all plausible loopholes to be closed.

Lubin, L. M.* & Sandage, A.[†] (June, 2001). The Tolman Surface Brightness Test for the Reality of the Expansion. IV. A Measurement of the Tolman Signal and the Luminosity Evolution of Early-Type Galaxies.

Astrophys. J., 517, 565-86; https://arxiv.org/pdf/astro-ph/0106566.

* Department of Astronomy, *California Institute of Technology, Pasadena*, California. Current address: Department of Physics and Astronomy, Johns Hopkins University, Baltimore, MD.

[†] *Observatories of the Carnegie Institution of Washington*, 813 Santa Barbara Street, *Pasadena*, California.

This article was included because it was one of very few recent "mainstream" articles which claimed to evaluate *Zwicky's "tired-light" theory* alongside tests for the *expansion of the universe*, but in fact it did not. It was very disappointing. It omitted an expansion factor, then failed to include the *"tired-light" factor* representing the linear loss of energy as the *photons* interact with *electrons* and other matter as they travel through intergalactic space. The evidence for an *expanding universe* which it provided based on *Tolman surface brightness test* was demonstrated by Lerner, E. J., Falomo, R., & Scarpa, R. (2014). *UV surface brightness of galaxies from the local universe to z ~ 5*, (below) to be in error, and, when calculated correctly, *concluded that far from disproving a non-expanding cosmology, data by Lubin and Sandage agreed very well with predictions for a static Euclidean universe*. However, this paper provides some useful insights on this analysis, and on the deeply ingrained biases at work.

At first reading I thought 'under which rock had these authors been living'. So, I investigated further. Allan Sandage (June 18, 1926 – November 13, 2010) was one of the most influential astronomers of the 20th century. He was born in Iowa City, Iowa, United States. He graduated from the University of Illinois in 1948. In 1953 he received a PhD from the *California Institute of Technology*; the German-born

Wilson Observatory-based astronomer Walter Baade was his advisor. During this time Sandage was a graduate student assistant to cosmologist Edwin Hubble. He continued Hubble's research program after Hubble died in 1953. In 1952 Baade surprised his fellow astronomers by announcing (at the 1952 Conference of the International Astronomical Union, in Rome) his determination of two separate populations of Cepheid variable stars in the Andromeda Galaxy, resulted in a doubling of the estimated age of the universe (from 1.8 to 3.6 billion years). Hubble had posited the earlier value; he had considered only the weaker Population II Cepheid variables as standard candles. After Baade's pronouncements, Sandage showed that astronomers' previous assumption, that the brightest stars in galaxies were of approximately equal inherent intensity, was mistaken in the case of H II regions which he found not to be stars and inherently brighter than the brightest stars in distant galaxies. This resulted in another 1.5-fold increase in the calculated age of the universe, to approximately 5.5 billion years. Throughout the 1950s and well into the 1980s Sandage was regarded as the pre-eminent observational cosmologist, making contributions to all aspects of the cosmological distance scale, ranging from calibrators within our own Milky Way Galaxy, to cosmologically distant galaxies. Sandage began working at the Palomar Observatory. In 1958 he published the first good estimate for the Hubble constant, revising Hubble's value of 250 down to 75 km/s/Mpc, which is close to today's accepted value. Later he became the chief advocate of an even lower value, around 50, corresponding to a Hubble age of around 20 billion years. At the time, many, especially Sandage, believed that the cosmological constant was zero.

Lori Lubin was an observational astronomer, who earned a Ph.D. in Astrophysical Sciences at Princeton University, 1995; was a Carnegie Postdoctoral Fellow, *Observatories of the Carnegie Institute of Washington*, from 1995-1997; a Hubble Postdoctoral Fellow, *California Institute of Technology*, from 1997-2000; and Associate Research Scientist, at *John Hopkins University*, 2000-2001, where she was when this article was published. When she published this article with Alan

Sandage, she would have been about 30, whilst Sandage was 75 years old. Following this collaboration, she was appointed Assistant Professor of Physics at the *University of California, Davis*, in 2002, Associate Professor of Physics, University of California, Davis, in 2004, and Professor of Physics, University of California, Davis, in 2009 until the present.

Abstract

We review a sample of the early literature in which the reality of the expansion is discussed. Hubble's reticence, even as late as 1953, to accept the expansion as real is explained as due to his use of equations for distances and absolute magnitudes of redshifted galaxies that do not conform to the modern *Mattig equations* of the standard model. [???]

[*Mattig's formula* was an important formula in observational cosmology and extragalactic astronomy which gives a relation between radial coordinate and redshift of a given source. *It depends on the cosmological model being used* and is used to calculate *luminosity distance* in terms of *redshift*. It assumes zero *dark energy*, and is therefore no longer applicable in modern cosmological models such as the *Lambda-CDM model*, (which require a numerical integration to get the distance-redshift relation). However, *Mattig's formula* was of considerable historical importance as the first analytic formula for the distance-redshift relationship for arbitrary matter density, and this spurred significant research in the 1960s and 1970s attempting to measure this relation.]

The *Tolman surface brightness test*, once the only known test for the reality of the expansion, is contrasted with three other modern tests.

[The *Tolman surface brightness test* is one out of six cosmological tests that were conceived in the 1930s to check the viability of and compare new cosmological models. *Tolman's test* compares the *surface brightness* of galaxies as a function of their *redshift* (measured as z). Such a comparison was first proposed in 1930 by Richard C. Tolman as a test of whether the universe is *expanding* or *static*. It is a unique test of cosmology, as it is independent of *dark energy*, *dark matter* and *Hubble constant* parameters, testing purely for whether *cosmological redshift* is caused by an *expanding universe* or not[1].

[1] Tolman, R. (1934). *Relativity Thermodynamics and Cosmology*. International Series of Monographs on Physics, Oxford.]

In a *simple (static and flat) universe*, the *light* received from an object drops proportional to the square of its *distance* and the *apparent area* of the object also drops proportional to the square of the *distance*, so the *surface brightness* (*light received* per [*apparent*] *surface area*) would be constant, independent of the distance. In an *expanding universe*, however, there are *two* effects that change this relation. First, the rate at which *photons* are received is reduced because each *photon* has to travel a little farther than the one before. Second, the *energy* of each *photon* observed is reduced by the *redshift* [z] ["tired-light" theory ???]. At the same time, *distant objects appear larger than they really are because the photons observed were emitted at a time when the object was closer* [*apparent area* of the object larger]. Adding these effects together, the *surface brightness* in a simple *expanding universe* (flat geometry and uniform expansion over the range of *redshifts* observed) should decrease with the fourth power of 1 + z, [where z is the redshift, ???].

One of the earliest and most comprehensive studies was published in 1996, as observational requirements limited the practicality of the test till then. This test found consistency with an *expanding universe*[2].

[2] Pahre, M. A., Djorgovski, S. G., & Carvalho, R. R. de (January, 1996). A Tolman Surface Brightness Test for Universal Expansion and the Evolution of Elliptical Galaxies in Distant Clusters. *The Astrophysical Journal*, 456, 2, L79; arXiv:astro-ph/9511061; https://doi.org/10.1086/ 309872.

However, therein, the authors note that: "The results of any Tolman *surface brightness* (SB) test where galaxies must be corrected to a standard condition will involve some dependence on the *assumed cosmology*, but as will be described below, for the *redshifts* of interest here, the effect of cosmology is quite small compared to the predicted difference between the *expansion* and *tired-light* models."

A later paper that reviewed this one *removed their assumed expansion cosmology* for calculating *surface brightness* (SB), to make for a fair test [!!!], and found that the 1996 results, once the correction was made, *did not rule out* a *static universe*[3].

[It did more than that. It concluded "that available observations of galactic *surface brightness* (SB) are consistent with a *static Euclidean universe* (SEU) model. See article below.]

[3] Lerner, E. J., Falomo, R., & Scarpa, R. (2014). UV surface brightness of galaxies from the local universe to z ~ 5. *International Journal of Modern Physics* D, 23, 6, 1450058; arXiv:1405.0275; https://doi.org/10.1142/ S0218271814500588.

To date, the most complex investigation of the relationship between *surface brightness* and *redshift* was carried out using the 10m Keck telescope to measure nearly a thousand galaxies' *redshifts* and the 2.4m Hubble Space Telescope to measure those galaxies' *surface brightness*[4].

[4] Sandage, A. & Lubin, L. M., 2271-88; Lori M. Lubin, L. M. & Sandage, A. (2001). The Tolman Surface Brightness Test for the Reality of the Expansion. II. The Effect of the Point-Spread Function and Galaxy Ellipticity on the Derived Photometric Parameters. *Astronomical Journal*, 121, 2289-300; Lori M. Lubin, L. M. & Sandage, A. (2001). The Tolman Surface Brightness Test for the Reality of the Expansion. III. Hubble Space Telescope Profile and Surface Brightness Data for Early-Type Galaxies in Three High-Redshift Clusters. *Astronomical Journal*, 122, 1071-83; Lori M. Lubin, L. M. & Sandage, A. (2001). The Tolman Surface Brightness Test for the Reality of the Expansion. IV. A Measurement of the Tolman Signal and the Luminosity Evolution of Early-Type Galaxies," *Astronomical Journal*, 122, 1084-103.

The exponent found is *not* 4 as expected in the simplest expanding model, but 2.6 or 3.4, depending on the frequency band. The authors summarize:

"We show that this is precisely the range expected from the *evolutionary models of Bruzual & Charlot*. We conclude that the *Tolman surface brightness test* is consistent with the reality of the expansion." [???]

[Bruzual, G., & Charlot, S. (October, 2003). Stellar population synthesis at the resolution of 2003. *Monthly Notices of the Royal Astronomical Society*, 344, 4, 1000-28; https://doi.org/10.1046/ j.1365-8711.2003.06897.x; https://doi.org/10.48550/arXiv.astro-ph/0309134: "We present a new model for computing the *spectral evolution* of *stellar* populations at ages between 1×10^5 and 2×10^{10} yr at a resolution of 3 Å across the whole *wavelength* range from 3200 to 9500 Å for a wide range of metallicities. These predictions are based on a newly available library of *observed stellar spectra*. We also compute the spectral evolution across a larger *wavelength* range, from 91 Å to 160 μm, at lower resolution. The model incorporates recent progress in *stellar evolution theory* and an observationally motivated prescription for thermally pulsing stars on the asymptotic giant branch. The latter is supported by observations of *surface brightness fluctuations* in nearby stellar populations. We show that this model reproduces well the observed optical and near-infrared color-magnitude diagrams of Galactic star clusters of various ages and metallicities.]

Some proceeding work has pointed out that the analysis tested *one possible static cosmology (analogous to Einstein–de Sitter), and that static models with different angular size-distance relationships can pass this test,*[3, 5].

[5] Lerner, E. J. (July, 2018). Observations contradict galaxy size and surface brightness predictions that are based on the expanding universe hypothesis. *Monthly Notices of the Royal Astronomical Society*, 477, 3, 3185–96; arXiv:1803.08382; https://doi.org/10.1093/mnras/sty728.

The predicted difference between *static* and *expansion* diverges dramatically towards higher *redshifts*, however, *accounting for galaxy evolution becomes increasingly uncertain*. The broadest test done to date was out to $z = 5$, this test found their results to be consistent with a *static universe*, but was unable to rule out expansion as it tested only a single model of galaxy size evolution[5]. *Static "tired-light" models* remain in conflict with observations of *supernovae*, as these models do not predict *cosmological time dilation*[6,7,8] [???]

[6] Errors in Tired-Light Cosmology. www.wolff.ch. Retrieved September 9, 2024.

[7] Goldhaber, G., Groom, D. E., Kim, A., Aldering, G., *et al.* (September, 2001). Timescale Stretch Parameterization of Type Ia Supernova B-Band Light Curves. *Astrophys. J.*, 558, 1, 359–68; arXiv:astro-ph/0104382; https://doi.org/10.1086/

322460.

[8] White, R. M. T., Davis, T. M., Lewis, G. F., Brout, D., Galbany, L., *et al.* (September, 2024). The Dark Energy Survey Supernova Program: slow supernovae show cosmological time dilation out to z 1". *Monthly Notices of the Royal Astronomical Society*, 533, 3, 3365–78; arXiv:2406.05050; https://doi.org/10.1093/mnras/stae2008. ISSN 0035-8711.

[*Cosmological time dilation* (or *cosmic time dilation*) is the phenomenon that events observed at cosmological distances (i.e., with a significant *cosmological redshift*) appear to us to take longer than if the event was nearby. Given the assumption of an *expanding distance* between us and the event (*Hubble expansion*), light takes increasing amounts of time to reach us over the course of the event. The lengthening ratio is 1 to 1 + z, the same ratio as *redshifted wavelengths* from the same distance, *and the two can be considered the same phenomenon.*

Cosmological time dilation would occur from the increasing distance irrespective of the *time dilation* associated with *relativity*, but *relativity* does affect the amount of *time dilation, significantly at cosmological distances (at which the receding motion is at a relativistic speed*); *special relativity* is relevant and the *Lorentz transformation* applies: the adjustment identical to the method used for a calculating a *redshifted* wave period. *General relativity* includes *time-dilation* effects due to gravity, and also has effects (beyond those of *special relativity*) that apply to distant objects.]].

These are (1) the *time dilation* in *Type Ia supernovae* light curves, (2) the temperature of the *relic radiation* as a function of *redshift*, and (3) the *surface brightness* normalization of the Planckian shape of the *relic radiation*.

We search for the *Tolman surface brightness depression* with *redshift* using the *Hubble Space Telescope* (HST) data from Paper III for 34 early-type galaxies from the three clusters Cl 1324+3011 (z = 0.76), Cl 1604+4304 (z = 0.90), and Cl 1604+4321 (z = 0.92). Depressions of the *surface brightness* relative to the *zero-redshift* fiducial lines in the mean *surface brightness*, log linear radius diagrams of Paper I are found for all three clusters. Expressed as the exponent, n, in 2.5 log (1 + z)n mag, the value of n averaged over Petrosian radii of η = 1.7 and η = 2.0 for all three clusters is n = 2.59 $\pm$ 0.17 in the R band and 3.37 $\pm$ 0.13 in the I band for a q_0 = 1/2 model. The sensitivity of the result to the assumed value of q_0 is shown to be less than 23% between q_0 = 0 and +1. The conclusion is that the exponent on (1 + z) varies from 2.28 to 2.81 ($\pm$ 0.17) in the R band and 3.06 to 3.55 ($\pm$ 0.13) in the I band, depending on the value of q_0. For a true Tolman signal with n = 4, the luminosity evolution in the look-back time, expressed as the exponent in 2.5 log $(1 + z)^{4-n}$ mag, must then be between 1.72 to 1.19 ($\pm$ 0.17) in the R band and 0.94 to 0.45 ($\pm$ 0.13) in the I band. We show that this is precisely the range expected from the evolutionary models of Bruzual & Charlot (1993, *Ap. J*, 405, 538) and other measurements of the luminosity evolution of early-type galaxies. *We conclude that the Tolman surface brightness test is consistent with the reality of the expansion to within the combined errors of the observed SB depression and the theoretical correction for luminosity evolution.* [???]

We have also used the *high-redshift Hubble Space Telescope* (HST) data to test the *"tired-light"* speculation [???] for a *non-expansion model* for the *redshift*. The HST data rule out the *"tired-light"* model at a significance level of better than 10 σ. [???]

---

1. Introduction

1.1. *Early Commentaries on the Reality of the Expansion*

With the announcement by Hubble [Hubble, E. (April, 1929). *A relation between distance and radial velocity among extra-galactic nebulae*, see

128

above] of a correlation between his estimates of distances to nearby galaxies and their redshifts, observational cosmology came of age, beginning its Long Journey into Night. The redshifts used by Hubble had been measured by Slipher, published by Eddington (1923), and added to by Humason (1929) in his crucial extension to higher values before Hubble's announcement.

However, the announcement of an *expanding universe* was such an extraordinary claim that proof of its reality by some independent means seemed essential, even though an *expanding universe* had been predicted by Friedmann (1922) [cf. Tropp, Frenkel & Chernin 1993] as one of the solutions of fundamental *Einstein equation of general relativity*. Expanding solutions were also detailed later by Lemaitre (1927, 1931) and Robertson (1928), each of which, acknowledging Friedmann, made advances beyond Friedmann by their adumbrations concerning the available observations of galaxies and their relative distances.

It is not clear whether Hubble knew of the Friedmann (1922) prediction or of the theoretical, cum observational, papers of Lemaitre and Robertson in 1927 and 1928, although Robertson told one of us (AS) that he had discussed with Hubble the existence of an expanding solution to the Einstein equations before 1929 (e.g. Sandage 1995, footnote 16 to Chapter 5). None of these three principal theoreticians are mentioned in Hubble's 1929 announcement.

Nevertheless, the decade of the 1920's was not entirely free of observational attempts to test a related cosmological prediction. A year after the publication of *Einstein's field equations*, de Sitter (1917a,b) had discovered one curious solution to them. Although his solution was that of a *static* metric (there are only three such solutions; Tolman 1929), it remarkably exhibited a *redshift*. The metric coefficient of the four-space time coordinate was a function of distance from the observer. Clocks that are further from the observer's origin in the three-space manifold would appear to tick more slowly than clocks at that origin. This effect would give an apparent *redshift* that would vary with distance. The formal feature of a *redshift* in the de Sitter metric was called *"the de Sitter effect."* A curious additional feature was that any test particle put into the de Sitter manifold would exhibit a radial motion (Eddington 1923, §70; de Sitter 1933; Tolman 1934, §144–145)[3] even though the spatial metric was *static*.

[3] An accessible derivation of the de Sitter static metric and its properties is given by Tolman (1934, §136 and §142–145). A more recent summary of the de Sitter metric with its redshift properties is given elsewhere (Sandage 1975, §2.2).

The predicted existence of the *de Sitter redshift effect* with its *static* metric became well known in the decade of the 1920's. Many attempts were made to find the effect using astronomical data (distances and velocities) for objects such as stars and globular clusters, thought then to comprise the wider universe before Hubble (1925, 1926, 1929b) proved the existence of external galaxies. Among the most accessible papers concerning the search for the de Sitter effect are those by Silberstein (1924), Stromberg (1925), Wirtz (1925), Lundmark (1925), and undoubtedly many others.

With regard to the *de Sitter effect*, it is a continuing curiosity as to what Hubble meant in the final sentence of his announcement in 1929 where he wrote: "The outstanding feature is the possibility that the *velocity-distance* relation may represent the *de Sitter effect*.... In this connection it may be emphasized that the linear relation found in the present discussion is a first approximation representing a restricted range in distance (emphasis added)".

In other words, a larger range in distance may not show a linear relation, he surmised, where he apparently knew that the first order *de Sitter effect* is quadratic in distance rather than linear (e.g. Sandage 1995, eq. 5.10). Cautiously, Hubble left open the possibility that the data which he had discussed might define a *redshift-distance* relation that would actually vary as the square of the distance, which, in first approximation, would appear to be linear for short distances near the origin (i.e. the first term of a Taylor series). Hubble also undoubtedly knew of the earlier searches for the *de Sitter effect* by Lundmark (1925) and Silberstein (1924) where they attempted parabolic fits to their adopted data.

However, Hubble & Humason (1931) soon proved that the *redshift-distance* relation was, in fact, linear over the much larger range of *redshifts* than was available in 1929. For this and other reasons, de Sitter (1933) wrote: "We know now, because of the observed expansion, that the actual universe must correspond to one of the non-static models.... The *static models* are, so to say, only of academic interest."

1.2. *Later Attempts to Counter the Reality of the Expansion*

Nevertheless, the concept of an *expanding universe* seemed so bizarre to many commentators that *attempts began already* in 1929 to find alternate ways to produce large *redshifts* other than from a true expansion. These attempts continue to this day.

The first alternate suggestion was made by Zwicky (1929) where he proposed that *photons* lose *energy* on their way to us from a distant source. [Zwicky, F. (1929). *On the Red Shift of Spectral Lines through Interstellar Space*, see above.]. As a consequence, they would show a *redshift* in their energy distribution, both in the continuum radiation and in the Fraunhofer lines used to measure the *redshifts*. To first order, the *redshift* effect would be linear with distance because the first term in a Taylor expansion of $1 + z = e^{HD/c}$ is linear in D. Here, H is the Hubble constant and D is the distance. With this suggestion, Zwicky (1929) introduced the notion of *"tired-light"*.

Even as late as mid-twentieth century, Zwicky (1957) maintained that the hypothesis was viable. However, *neither Zwicky nor any other subsequent supporter of the proposition (e.g. La Violette 1986; Pecker & Vigier 1987) gave a convincing physical theory for the "tiredness"*. As critics still point out, any *scattering process* with *energy* transfer from the photon beam to the scattering medium, as required for a *redshift*, must broaden (deflect) the beam. This effect would cause images of distant galaxies to be fuzzier than their local counterparts, which they are not.

> [As Zwicky made clear he did not propose a "*scattering process*" but a *transfer of energy* from the interaction between the photons and the electrons ore other matter: "it is evident that *any explanation based on a scattering process like the Compton effect or the Raman effect, etc., will be in a hopeless position regarding the good definition of the images as mentioned under B₄.*"
>
> ... "a light quantum ħv has an *inertial* and a *gravitational mass* ħv/c². It should be expected, therefore, that a quantum ħv passing a mass M will not only be deflected but it will also transfer *momentum* and *energy* to the mass M and make it recoil. During this process, the light quantum will change its *energy* and, therefore, its *frequency*."]

1.3. *Hubble's Reticence*

Perhaps the most interesting attack on the reality of the expansion was *the reluctance of Hubble himself to believe that the redshifts represent a true expansion* rather than "caused by an unknown law of nature". Much has been written about Hubble's reluctance, most of which is wrong. Some commentators even suggest philosophical or religious reasons related to a presumed abhorrence of a "creation" event that is implied in some interpretations of a real Friedmann expansion.

However, the fact is that Hubble's reasons were those of a reductionist bench scientist. He relied (mistakenly, it turns out) solely on the interpretation of his observational data and their accuracy, coupled with a *mistaken theory* of how *redshifts* should vary with "distance." [???] His equation for "distance" has no justification *within the modern [relativistic] Mattig (1958) equations*. This story, and why Hubble's conclusion would not have been reached using current data analyzed with the modern theoretical equations, is set out in detail elsewhere (Sandage 1998).

In outline, Hubble's argument was as follows. By the mid-1930's, Hubble (1934; 1936a,b) had completed his program of galaxy counts, the goal of which was to measure the curvature of space. He also completed the extension of the Hubble diagram of *redshift* versus *apparent magnitude* to the limit of the Mount Wilson 100-inch reflector (Hubble 1936; Humason 1936). The data for each of these programs had to be corrected for the effects of *redshifts* on the *apparent magnitudes*. If the expansion was real, Hubble assumed that the observed bolometric magnitudes (or equivalently, the observed magnitudes corrected for the selective part of the K term) must be made brighter by two factors of 2.5 log $(1 + z)$; however, if the *redshift* was due to "*an unknown law of nature*" [???] rather than the expansion, only one such factor was to be applied to the observed magnitudes.

Hubble believed that the Hubble diagram (log of the *redshift* versus *apparent magnitude*) must be strictly linear; in addition, the radius of curvature of space implied by his "corrected" magnitudes for his galaxy counts must not be "too small." Consequently, he became convinced that only one factor of 2.5 log $(1 + z)$ should be applied. If so, *the expansion would not be real.* Hubble (1936b) wrote:

> "If the *redshifts* are not primarily due to velocity shifts ... [then] the velocity-distance relation is linear; the distribution of nebulae is uniform; *there is no evidence of expansion*, no trace of curvature, no restriction of the time scale.... The unexpected and truly remarkable features are introduced by the additional assumption that *redshifts* measure recession. The velocity-distance relation deviates from linearity by the exact amount of the postulated recession. The distribution departs from uniformity by the exact amount of the recession. The departures are compensated by curvature which is the exact equivalent of the recession. Unless the coincidences are evidence of an underlying necessary relation between the various factors, they detract materially from the plausibility of the interpretation. The small scale of the expanding model, both in space and time, is a novelty

and as such will require rather decisive evidence for its acceptance."

That "rather decisive evidence" is now available from at least three modern experiments that are independent of the *Tolman galaxy surface brightness test*. [???]

1.4. Other Recent Proofs that the Expansion is Real

1.4.1. *The Time Dilation Test*

The *Tolman (1930) surface brightness* $(1 + z)^4$ *effect* was the only known test for the reality of the expansion until Wilson (1939) suggested that *the shape of the light curves of Type Ia supernovae provides a clock*. This supposition was based on the uniform shape of the light curve discovered by Baade (1938) and recalled as history by Minkowski (1964). Wilson reasoned that such clocks at different *redshifts* would measure the *special relativity time dilation* if the light curve shapes of *SNe Ia* at high *redshifts* could be observed.

[Depends on Einstein's *theory of Special Relativity*. Does not apply under New Physics.]

The stretching of the light curves, increasing with *redshift*, has now been observed. The data give a spectacular confirmation of the *time dilation* effect (Goldhaber *et al.* 1997, 2001). [???]

1.4.2. *Temperature of the Relic Blackbody Radiation as a Function of Redshift*

Blackbody radiation in an expanding cavity remains Planckian in shape but with a decreasing temperature that scales as $T(z) = T(0)(1 + z)$ with *redshift* (Tolman 1934, eq. 177.7; Bahcall & Wolf 1968). The observations of the *Boltzman temperature* of *interstellar molecules* in the spectra of high-*redshift* galaxies has now apparently been measured in a difficult experiment with the Keck 10-m telescopes (Songaila *et al.* 1994). An important confirmation of these first results was made in observations by Ge *et al.* (1997) and Srianand, Petitjean & Ledoux (2000). Only upper limits had been achieved before (cf. Meyer *et al.* 1986) which, nevertheless, were highly important in pioneering this test.

1.4.3. *Measurement of the Chemical Potential the Alpher-Herman Relic Black Body Radiation*

Although the shape of an initial *blackbody spectrum* remains Planckian in an expanding cavity, the vertical normalization (i.e. the *photon number*) remains Planckian only if that normalization is decreased with *redshift* by $(1 + z)^4$. This fact is derived trivially from the *Planck equation* as expressed in terms of *energy flux* per *wavelength* per unit *wavelength* interval rather than *frequency* per unit *frequency* interval. (Of course, both representations are equivalent by minding the relation between *wavelength* interval and *frequency* interval, where $d\nu = cd\lambda/\lambda^2$.)

> [*Planck's law* (also *Planck radiation law*) describes the spectral density of *electromagnetic radiation* emitted by a *black body* in *thermal equilibrium* at a given temperature T, when there is no net flow of matter or energy between the body and its environment.
>
> At the end of the 19th century, physicists were unable to explain why the observed spectrum of *black-body radiation*, which by then had been accurately measured, diverged significantly at higher frequencies from that predicted by existing theories. In 1900, German physicist Max Planck heuristically derived a formula for the observed spectrum by assuming that a hypothetical electrically charged oscillator in a cavity that contained black-body radiation could only change its energy in a minimal increment, E, that was proportional to the frequency of its associated electromagnetic wave. While Planck originally regarded the hypothesis of dividing energy into increments as a mathematical artifice, introduced merely to get the correct answer, other physicists including Albert Einstein built on his

135

work, and Planck's insight is now recognized to be of fundamental importance to quantum theory.]

Hence, because the *Planck equation* defines a *surface brightness, a test of the Tolman surface brightness effect is equivalent to measuring the deviation of the photon number per unit surface area in the sky by comparing the observations with the normalization given by the Planck equation itself.* The deviation of the data from the Planck equation is called the *"chemical potential"*. Among other things, the deviation with *wavelength* could be due to *Compton scattering* in the early universe.

No deviation has been found in the observations to within one part in 10^4. The perfect Planckian shape of the relic radiation was measured with COBE to within this limit of 9×10^{-5} (Mather *et al.* 1990; Fixsen *et al.* 1996). The conclusion to be drawn from this spectacular result is that this seemingly perfect normalization of the spectral energy distribution *is a definitive proof of the Tolman surface brightness factor and, therefore, a definitive proof of the reality of the expansion.* [This is a stretch.] We shall adopt this assumption later in §4 where we combine our *surface brightness* signal with the theoretical $(1 + z)^4$ Tolman signal and interpret the difference to be due to *luminosity* evolution in the high-*redshift* look-back time. [???]

1.5. Plan of the Present Paper

In the first three papers of this series (Sandage & Lubin 2001, hereafter Paper I; Lubin & Sandage 2001a and 2001b, hereafter Papers II and III), we provide the background and observational data required to carry out the Tolman test. Most importantly, we measure the fiducial (zero *redshift*) relations between the *mean surface brightness, absolute magnitude,* and *linear radius* for local early-type galaxies in Paper I. In Paper III, we present the observational data on our high *redshift* comparison sample of 34 early-type galaxies in the three clusters used in this program, Cl 1324+3011 at z = 0.76, Cl 1604+4304 at z = 0.90, and Cl 1604+4321 at z = 0.92 (Oke, Postman & Lubin 1998; Postman *et al.* 1998, 2001; Lubin *et al.* 1998, 2001). To compare accurately the local and high-*redshift* data, the parameters of each galaxy are measured at discrete values of the Petrosian (1976) η *metric radius*, which is defined as the difference in magnitude between the *mean surface brightness* averaged over the area interior to a particular *radius* and the *surface brightness* at that *radius* (see §2 of Paper I).

In the present paper, we use the results presented in Papers I–III to complete the *Tolman test*. In §2 we review the *Mattig (1958) cosmological equations* with which to calculate the *absolute magnitudes* and *linear radii* of galaxies from their *apparent magnitudes* and *angular radii*. Explicit equations for the special case of $q_0 = 1/2$ are given. We use these equations to obtain the *total magnitude* M, *linear radius* R, and SB as functions of five Petrosian η radii of 1.0, *mean surface brightness* 1.3, 1.5, 1.7, and 2.0 mag for the 34 high-*redshift* early-type galaxies. Tables 2–4 list these values for $q_0 = 1/2$ and $H_0 = 50$ km s^{-1} Mpc^{-1}. These values are derived from the observational data listed in Paper III. The *Tolman test* is made in §3 based on the data in §2 and the comparison with the *surface brightness* data at zero *redshift* from Paper I.

In §4 we set out the theoretical evolutionary corrections, first, using the simplest model of passive evolution by *main sequence burn-down* in the HR diagram as a function of time and, second, using the sophisticated star-formation models of Bruzual & Charlot (1993), following earlier papers by Guiderdoni & Rocca-Volmerange (1987, 1988) and Rocca-Volmerange & Guiderdoni (1988).

In §5 we show the sensitivity of the observed Tolman signal to the assumed value of $q_0 = 1/2$ used in the data of Tables 2–4. *Proof that the "tired-light" assumption does not fit the data by a large factor* is given in §6. In §7, we discuss the systematic uncertainties in the *Tolman test* made here and describe plans to strengthen the present test.

2. Calculation of Absolute Magnitudes and Linear Radii for the High-Redshift Galaxies

Comparison of the high-*redshift* data in Paper III with the data for local galaxies in Paper I requires knowledge of *absolute magnitudes* (corrected for K term) and *linear radii*. We have two options to calculate these values.

(1) If we were to assume that the standard model that leads to the *Mattig* (1958, 1959) *equations* (*which assume that the expansion is real*) did not exist, the natural assumption is that the distance, D_0, at the time that light is received is related to *redshift* by $D_0 = cz/H_0$. This was Hubble's assumption throughout his work, including the last summary paper in his Darwin lecture (Hubble 1953).

(2) Alternately, we can adopt the details of the standard model by choosing a value for the *deceleration parameter*, q_0 (Hoyle & Sandage 1956). *We then use the Mattig (1958) equations* to calculate the *distance modulii*, m−M, and the *linear radii*, R, from the observed *angular radii*[4].

[4] To avoid confusion with our use of R in this paper to mean the *linear radius* of a galaxy rather than its use as the time-dependent *scale factor* in the metric, we have used D here as the *distance parameter* by replacing Rr by D, where, in the standard notation, R is the scale factor that varies with time and r is the dimensionless co-moving *radial coordinate* in the metric. Hence, in our notation here, D_0 is the distance at the time light is received from a galaxy at *redshift* z. *This is given by Mattig* (1958; see also Sandage 1988, eq. 30) as
$$D_0 = c/\{H_0 q_0^2(1 + z)\} \, [zq_0 + (q_0 - 1)\{-1 + \sqrt{(2q_0 z +1)}]. \quad (1)].$$

Because these equations already contain the Tolman $(1 + z)^4$ factor, the test for the *Tolman effect* becomes one of consistency between the *surface brightness* observations and the predictions of the *standard model*. The argument of why this does not lead to a hermeneutical circularity was given in §5 of Paper I. *We adopt option (2) in calculating the distance moduli and the linear radii of the program clusters.* We treat only the $q_0 = 1/2$ case of the standard model in this section. The cases for $q_0 = 0$ and +1 are treated in §5 using the correction recipes in Table 8.

2.1. The Equations for $q_0 = 1/2$

For the $q_0 = 1/2$ case with $H_0 = 50$ km s^{-1} Mpc^{-1}, the *Mattig equations* reduce to

$$m - M = 5 \log [2(1 + z - \sqrt{(1 + z)}] + 43.89, \qquad (2)$$

and

$$D_1 = 2c/\{H_0(1 + z) [1 - 1/\sqrt{(1 + z)}], \qquad (3)$$

where D_1 is the *distance* (in parsecs) when light left the galaxy at the observed *redshift* z. The *linear radius* R, in parsecs, of a galaxy with *angular radius* θ, in radians, is

$$R(pc) = 2c \{\sqrt{(1 + z)} - 1\}/H_0(1 + z)^{3/2} \theta \text{ (radians)}, \qquad (4)$$

using the distance D_1 from equation (3).

Changing the *angular radius* in radians to arcseconds, equation (4) can be conveniently written as

$$\log R(pc) = \log \theta'' + \log A, \qquad (5)$$

where

$$A = 5.818 \times 10^4 [\{\sqrt{(1 + z)} - 1\}/(1 + z)^{3/2}], \qquad (6)$$

for $q_0 = 1/2$ and $H_0 = 50$.

Equations 2–4 are derived in Sandage (1961a,b; 1988; 1995), where the *standard model* is summarized from the point of view of practical cosmology. Table 1 shows the $m - M$ and A values for the three program clusters for the $q_0 = 1/2$ case using equations (2), (5), and (6). The K terms are those derived in Paper III (see Table 4).

...

6. The Tired-Light Model Compared with the Observations

In contrast with the *standard model* with the *Mattig equations*, there is no metric theory of how distances and magnitudes are measured in a *"tired-light"* model. Therefore, we must guess at a reasonable equation for distance. We adopt a discussion that is given elsewhere (Sandage 1995, §4.3) and use *an equation for "coordinate" distance* in a *flat space* which is

$$D = c/H_0 \ln(1 + z). \tag{12}$$

Because the universe is not expanding in the model, the *distance "now"* in equation (12) *is also the distance when light left*. This, of course, is the crucial point. In the expanding case, the distance at the present epoch must be divided by $(1 + z)$ to give the distance when light was emitted. It is this latter distance in the expanding case that, when multiplied by the *angular radius*, fixes the *linear radius*. With $H_0 = 50$, the *linear radius* that corresponds to an observed *angular radius* (in arcsec) in the *"tired-light"* case is

$$\log R(pc) = 4.464 + \log [\ln(1 + z)] + \log \theta", \tag{13}$$

using equation (12). Hence, the A term, as defined in equation (5),
[For the $q_0 = 1/2$ case with $H_0 = 50$ km s^{-1} Mpc^{-1}, the *Mattig equations* reduce to

$$\log R(pc) = \log \theta" + \log A, \tag{5}$$

where

$$A = 5.818 \times 10^4 [\{\sqrt{(1 + z)} - 1\}/(1 + z)^{3/2}], \tag{6}$$

for $q_0 = 1/2$ and $H_0 = 50$]

becomes

$$\log A = 4.464 + \log [\ln(1 + z)]. \tag{14}$$

[This is not the *"tired-light"* theory, in which the *redshift* is caused by *photons* losing *energy* to *electrons* and other matter in intergalactic space.]

The log A values calculated in this way are listed in column 12 of Table 8 for the three high *redshift* clusters. These values, compared with the log A values for the $q_0 = 1/2$ case, give the increase in the log R (*linear radius*) values that must be entered in the *surface brightness* (SB), log R diagnostic diagram. For example, for Cl 1324+3011 (z = 0.7565) the log R values must be made larger by 0.305 dex relative to the $q_0 = 1/2$ values listed in Table 3.

The magnitudes must also be changed because the *distance* moduli are different than those in the fiducial $q_0 = 1/2$ case. The *absolute magnitude* calculation for the *tired-light model* follows from the expected theoretical relation that

$$l = L/4\pi D^2(1 + z), \tag{15}$$

where only one power of $(1 + z)$ for the "*energy effect*" is required, rather than two powers of $(1 + z)$ in the Roberston (1938) equation in the *standard theory* (cf. Sandage 1995, eq. 2.1). Hence,

$$m - M = 2.5 \log (1 + z) + 5 \log [\ln(1 + z)] + 43.89, \tag{16}$$

for $H_0 = 50$.

> [However, *there is an additional factor* in the "*tired-light*" case, representing the linear loss of *energy* as the *photons* interact with *electrons* and other matter as they travel through intergalactic space.]

It is can be shown from equation (16) that the *magnitude–redshift relation* for "*tired-light*" is the same to within a few hundredths of a magnitude with the case for $q_0 = +1$ using the *Mattig equation*.

> [But this does not represent the "*tired-light*" theory. It does not include the *factor* representing the linear loss of *energy* as the *photons* interact with *electrons* and other matter as they travel through intergalactic space.]

This is true even for *redshifts* as large as z = 1. Column 13 of Table 8 confirms this statement, seen by the fact that the entries in column 9 for the q_0 = +1 case are very close to those in column 13 for "*tired-light*". However, these magnitude changes are academic here in using the *surface brightness* (SB), log R diagnostic diagram because no corrections for the *magnitude* differences need to be applied to the SB values. These are observed SB values (only corrected for K dimming); hence, as emphasized before, they are independent of all cosmologies. With the changes to Tables 2–4 for log R that are listed in column 14 of Table 8, we can enter Figures 1 and 2 for the "*tired-light*" case in order to measure the depression from the local upper envelope calibration. The result, not shown but which the reader can recover using the Δ log R values given in Table 8 together with Tables 2–4, is that a measurable SB depression at these larger log R values is again present. It is, of course, smaller than for the expanding case because of the larger log R values. But, by how much? The crucial question is whether it is so much smaller as to conform to a depression of only 2.5 log (1 + z) mag when the correction for *luminosity* evolution is also applied.

We have analyzed the data for the three clusters in the same way as in §3; however, *we now use the correct log R values required by the "tired-light" conjecture*, using the recipes in Table 8. Expressing the result of the *surface brightness* (SB) depression from the *zero-redshift* fiducial line in the SB, log R diagram as the exponent, n, in 2.5 log $(1 + z)^n$ gives the weighted mean of n = 1.61 ± 0.13 for the R band and n = 2.27 ± 0.12 for the I band. As described in §3, we have used the results from the most reliable η values of η = 1.7 and 2.0. The resulting exponents are too large by approximately 5 and 10 σ, respectively, compared to the n = 1 prediction of the "*tired-light*" scenario. Consequently, to produce coherence with the "*tired-light*" model, we require negative *luminosity* evolution in the look-back time, i.e. galaxies must be fainter in the past. No feasible model of stellar evolution can produce such *luminosity* evolution with time.

In fact, just as in the expanding case, positive *luminosity* evolution must have occurred in the look-back time because there is no way in the *tired-light model* to prevent the stellar content of galaxies from evolving during the look-back time.

A *static model* where the *redshift* is not due to expansion is not the same as a *steady state model* where the *mean parameters of galaxies, averaged over an ensemble of galaxies, are required to be the same at all distances and at all times*. For a *steady state* to exist, despite the evolution of the stellar content of individual galaxies, requires that there must be young and old galaxies in every volume of space and at every cosmic time such that the mean age is the same at all distances and times. This requires continuous galaxy formation at the same rate at all cosmic times in order to maintain a constant mean age everywhere, always.

However, such *steady state models* are not the same as *static models* where the *redshift* is due to an *unknown* physical cause, either at the source or in the intervening light path from the source to the observer as originally postulated by Zwicky (1929). In fact, the *steady state models* proposed by Bondi, Gold, and Hoyle are truly *expanding models* where the *redshift* is due to the expansion.

> [This is not the case of Zwicky's "tired-light" theory, where the *redshift* is not due to an "unknown physical cause".]

They are not *static models*. Furthermore, an early proof that *steady state models* cannot be correct was the demonstration of the failure of the predicted *steady state* color distribution, with its required mixture of ages, to match the observed color distribution of early-type galaxies (Sandage 1973).

Hence, as in the present paper, *evolutionary corrections to magnitudes must be applied to both the expanding and the static models in making the present test*. The supposed degeneracy of the *Tolman test* due to the identity claimed by Moles *et al.* (1998) of the *surface brightness effect* in both the expanding and a *static (tired-light)* case is not correct. Their

143

error is due to a category mistake by confusing *static* and *steady-state* models. What Moles *et al.* have done is to combine a *static model* with a *steady state model*. This is a higher order departure from the standard *expanding model* than we have considered here and is not the test we have made. In any case, as stated above, a *steady state model* can again be disproved by the color argument (Sandage 1973). Hence, even the higher order model proposed by Moles *et al.* cannot be correct on this ground alone.

The result of the present paper is that a *static model*, where the redshift is due to an *unknown* physical cause, fails the *surface brightness test* by a large factor.

[But this is not true of Zwicky's *"tired-light"* theory.]

Such a model requires *luminosity evolution* in the look-back time, just as in the expanding case. Based on the analysis in §4, we find that a good approximation for the amount of increase in *luminosity* at the epoch of light emission is 2.5 log $(1 + z)^p$ mag, where p > 0.7. Applying this correction to the *"tired-light"* analysis gives the intrinsic "tired-light" prediction for the corrected exponents of > 2.31 (± 0.13) and > 2.87 (± 0.12) for the R and I bands, respectively. Each value is more than 10 σ from the required exponent of 1.0 if the *"tired-light"* scenario were correct.

[But this does not apply to Zwicky's *"tired-light"* theory.]

We take this to be a definitive proof that the hypothesis of non-expansion does not fit the surface brightness data [???].

7. Systematic Uncertainties in the Experiment: How can the Present Result be Improved?

There are *two systematic uncertainties* in the present experiment. Although neither of them is severe enough to jeopardize the results presented in §3 and §5 that the expansion is real, each can be overcome by more data with expanded boundary conditions on the parameters compared to the data used here. First, *the principal uncertainty at small radii (log R < 4.0) is the position of the zero-redshift fiducial line in the SB, log R diagram, relative to which the SB depressions for high-redshift galaxies are compared.* The Postman & Lauer (1995) data (Figure 2 of Paper I) *do not extend to radii smaller than log R(pc) = 4.0. Their data are confined to the first ranked cluster early type galaxies.* They do not sample into the *luminosity function* of each cluster to provide data for smaller galaxies. We have extended the data to smaller radii with the sample in Sandage & Perelmuter (1991) to generate a non-linear correction at small radii to the best-fit linear equations to the Postman & Lauer (1995) data (see Table 3 of Paper I). However, the Sandage & Perelmuter (1991) data *are also only for the first few, brightest cluster galaxies, again not going far into the fainter part of the luminosity function.* Hence, although Table 3 of Paper I gives our adopted extension to radii as small as log R(pc) = 3.3, *the uncertainties are large* and can be reduced by a more complete study of the SB, log R relation for fainter and smaller galaxies at low *redshift.*

Second, *the three clusters studied here are near the faint end of the distribution of absolute magnitude of first ranked galaxies* in, for example, the sample of first ranked cluster galaxies whose data are listed by Kristian, Sandage & Westphal (1978). The mean of the distribution of *absolute magnitude* in the R band for the Kristian *et al.* sample is MR = −24.5, whereas the first-ranked early-type galaxies in the three clusters studied have *absolute magnitudes* of MR = −23.7 for Cl 1604+4321, −23.7 for Cl 1604+4304, and −24.2 for Cl 1324+3011 (see PLO01). (Note again that the system of RJ magnitudes used by Kristian *et al.* is 0.25 mag brighter than the Cape R system used here). *The galaxies in the three clusters studied here have fainter absolute magnitudes and smaller radii than the average local clusters, exacerbating the problem described above.* Richer *high-redshift* clusters with brighter and, therefore, larger first-ranked galaxies *are known.* A study of the *Hubble Space Telescope* (HST) data from such clusters will improve the present *Tolman test.* We suspect that the present experiment is only the beginning of similar work

145

that will be done in the coming years with HST on such clusters as they are discovered and observed.

We are grateful to Mark Postman and J. B. Oke for their permission to use part of their extensive Keck and HST database for this paper and the previous papers in the series. LML would like to thank Chris Fassnacht for his essential material aids to this paper. AS is grateful for a conversation with James Peebles concerning the proof of the Tolman $(1 + z)^4$ factor from the exact normalization of the observed Planck black body background radiation in the COBE experiment, as discussed in §1.4.3.

LML was supported through Hubble Fellowship grant HF-01095.01-97A from the Space Telescope Science Institute, which is operated by the Associated Universities for Research in Astronomy, Inc. under NASA contract NAS 5-26555. AS acknowledges support for publication from NASA grants GO-5427.01-93A and G-06459.01-95A for work that is related to data taken with the Hubble Space Telescope

146

Seife, C. (June, 2001). 'Tired-Light' Hypothesis Gets Re-Tired.

Science, 292, 5526, 2414; https://www.science.org/doi/10.1126/science.292.5526.2414a.

Charles Siefe (journalist).

The *"tired-light" hypothesis*, mainstay of a dwindling band of contrarians who deny the *big bang* and its corollary, the *expanding universe*, has suffered a one-two punch. Observations of *supernovae* and of galaxies provide the best direct evidence that the universe is truly expanding and promise to shed light on the evolution of galaxies to boot.

"The expansion is real. It's not due to an unknown physical process. That is the conclusion," says Allan Sandage, an astrophysicist at the Carnegie Observatories in Pasadena, California, and leader of the galaxy study.

It's a conclusion that most astronomers reached long ago. In 1929, Edwin Hubble announced that light from distant galaxies is redder than light from nearby ones. Hubble and others took the redshifts as evidence that the universe is expanding [???], causing distant galaxies to speed away faster than nearby ones. To an observer on Earth, they reasoned, this would appear to stretch the wavelength of their light, just as the sound of a police-car siren seems to drop in frequency as it speeds away. However, within a few months of the publication of Hubble's paper, astrophysicist Franz Zwicky came up with an alternative explanation: that galaxies' light reddens because it loses energy as it passes through space. In Zwicky's *tired-light scenario*, the universe doesn't expand at all. Distant galaxies are red not because they are moving, but because their light has traveled farther and gotten pooped along the way.

"Tired-light"—a radical alternative to the standard expanding-universe model of the cosmos—has just failed two crucial tests.

When experimenters first measured the cosmic microwave background more than 30 years ago, they found that the radiation was too dim to be explained by Zwicky's hypothesis. That realization relegated *"tired-light"* firmly to the fringe of physics, but scientists still sought more direct proofs of the expansion of the cosmos.

Two new papers provide the best direct evidence yet.

[Both depend on *relativistic* assumptions. Does not apply in New Physics.]

The first, slated to appear in *Astrophysical Journal*, measures the brightening and dimming of a certain type of *supernova*[1].

[1] Perlmutter, S.*, Aldering, G., Goldhaber, G.*, Knop, R.A., Nugent, P., *et al.* (December, 1998). Measurements of Omega and Lambda from 42 High-Redshift Supernovae. *Astrophys. J.*, 517, 565-86; https://arxiv.org/pdf/astro-ph/9812133. See below.

Thanks to *Einstein's theory of relativity*, if distant *supernovae* are speeding away from us, they will appear to flare and fade at a more leisurely pace than close-by ones. A team of scientists led by Gerson Goldhaber of the Lawrence Berkeley National Laboratory (LBNL) in Berkeley, California, has shown that this is, indeed, the case with 42 recently analyzed *supernovae*. "It's such a clean-looking curve," says Saul Perlmutter, a member of the LBNL team. "It's very unambiguous."

In the second study, Sandage and Lori Lubin of Johns Hopkins University in Baltimore analyzed space-based measurements of the *surface brightness* of galaxies.

[2] Lubin, L. M. & Sandage, A. (June, 2001). The Tolman Surface Brightness Test for the Reality of the Expansion. IV. A Measurement of the Tolman Signal and the Luminosity Evolution of Early-Type Galaxies. *Astrophys. J.*, 517:565-586,1999; https://arxiv.org/pdf/astro-ph/0106566.

Both the *standard expanding-universe* and the *tired-light theory*, they realized, agree that redshifted light should make distant galaxies look dimmer than they really are. In an *expanding universe*, however, *time dilation* and other *relativistic* distortions will also dim distant galaxies, making them appear much fainter than *tired-light theory* dictates. What's more, young stars—and thus young galaxies—tend to be considerably brighter than old ones. When that extra brightness is taken into account, the observations match *expanding-universe* predictions, as Lubin and Sandage will report in *Astronomical Journal*. For the *tired-light theory* to be correct, young galaxies would have to be dimmer, rather than brighter, than old ones. [Not true.] "There's no way to explain that," says Lubin.

Although not surprising in themselves, the results are useful for "tidying things up in our cosmology," says Michael Pahre, an astronomer at the Harvard-Smithsonian Center for Astrophysics in Cambridge, Massachusetts, who performed a similar surface-brightness experiment in the mid-1990s. By comparing the *expanding-universe* theory's predictions with observed values of the *surface brightness* of distant galaxies, scientists can work backward and figure out how much brighter those galaxies must have been earlier in the history of the universe.

Even so, researchers doubt whether the results will convert tired-light diehards. "I don't think it's possible to convince people who are holding on to tired-light," says Ned Wright, an astrophysicist at the University of California, Los Angeles. "I would say it is more a problem for a psychological journal than for *Science*."

Mamas, D. L. (2010). An explanation for the cosmological redshift.

Phys. Essays, 23, 326; https://tiredlight.net/wp-content/uploads/2014/09/physics-essays-2010.pdf.

PhD Physics (UCLA). 4415 Clwr. Hr. Dr. N., Largo, Florida 33770.

Received: February 20, 2009.
Accepted: March 27, 2010.
Published online: April 29, 2010.

This is a bizarre attempt by an unrecognized author based in Florida (with a PhD in physics from UCLA) to reformulate *Zwicky's "tired-light" theory* in terms of a simplistic pseudo-classical notion of a *photon* as an *electromagnetic wave* which causes a *free electron* to oscillate and reradiate. It is included here as it provides useful background in the absence of any mainstream paper. It also provides calculations that suggest that the *cosmological redshift*, based on the quantum mechanical formulation of *Zwicky's "tired-light" theory* and estimates of the density of *free electrons* in intergalactic space, support a *static* rather than an *expanding universe*, on which the *Big Bang* origin of the universe is based. It incorrectly assumes, as Zwicky stated, that *Compton scattering* suffers from the blurring of images.

Abstract: A new theoretical model is presented which accounts for the *cosmological redshift* in a *static universe*. In this model the *photon* is viewed as an *electromagnetic wave* whose *electric field* component causes oscillations in deep space *free electrons* which then reradiate *energy* from the *photon*, causing a *redshift*. The predicted *redshift* coincides with the data of the Hubble diagram. The predicted *redshift* expression allows for the first time distance measurements to the furthest observable objects, without having to rely on their apparent magnitudes which may be subject to cosmic dust. This new theoretical model is not the same as, and is *fundamentally different from, Compton scattering*, and therefore avoids any problems associated with *Compton scattering* such as the blurring of images.

I. INTRODUCTION

Compton scattering has long been rejected as an explanation for the *cosmological redshift* because in this *particle-particle interaction, photons* are scattered into various angles at various frequencies, resulting in a blurring of images[1]. [Zwicky stated this, but a closer reading of Compton (May, 1923), shows that this was incorrect.]

[1] Zwicky, F. (1929). *Proc. Natl. Acad. Sci.*, 15, 773.

Numerous other mechanisms have been attempted to explain the *cosmological redshift*, such as an energy loss of the photon when traversing a radiation field[2], an inelastic scattering by gaseous atoms and molecules[3], or a dispersive-extinction effect by the space medium[4,5].

[2] Finlay-Freundlich, E. (1954). Red-Shifts in the Spectra of Celestial Bodies. *Proc. Phys. Soc., London*, A, 67, 192; https://doi.org/10.1088/0370-1298/67/2/114.
[3] Marmet, P. (1988). *Phys. Essays*, 1, 24.
[4] Wang, L. J. (2005). *Phys. Essays*, 18, 177.
[5] Wang, L. J. (2008). *Phys. Essays*, 21, 233.

Previously unconsidered by the principle of complementarity, a *photon* may also be viewed as a wave, interacting with intergalactic free electrons in a *wave-particle* fashion. It is reasonable to assume that although very short wave length *photons* (gamma rays) can interact with *electrons* in a *particle-particle* fashion (*Compton scattering*), *photons* of longer wavelengths than those of *gamma rays* could interact with *electrons* in a *wave-particle* fashion, the *electron* reacting to the *photon*'s *electric field*. [???]. Being that a *wavelength* of *visible light* is eight orders of magnitude larger than an *electron*, a visible wavelength *photon* should pass directly over an *electron* with unchanging direction and with

151

negligible blurring of images. This would circumvent Zwicky's above mentioned historical objection to *Compton scattering* over blurring of images and satisfy the consideration that *photons* travel *without appreciable transverse deflection*[6].

[6] Hubble, E. & Tolman, R. C. (1935). *Astrophys. J.*, 82, 302.

Furthermore, any other objection to *Compton scattering* (particle-particle) as an explanation for the *cosmological redshift* is irrelevant to the thesis of this present article, which does not propose a *Compton scattering* explanation but rather a fundamentally different *redshift mechanism* based on an instead *wave-particle interaction*.

The thesis of this present article is a theoretical prediction of a new mechanism, a new fashion in which a *photon* could interact with *free electrons* in deep space. This new theoretical model is supported by the calculations provided below, the predicted *cosmological redshift* coinciding with the data of the Hubble diagram.

A clear distinction is being drawn here between the case of extremely high frequency (*gamma ray*) *Compton scattering* (particle-particle) interactions, and the different manner in which *photons* of longer wavelengths than those of *gamma rays* may interact with free *electrons* in an instead *wave-particle* fashion.

> [This is not necessary, nor is a new mechanism. This is already explained in quantum theoretical terms by the interaction of *photons* with *electrons* and other matter in *intergalactic space*, in effect by a quantum mechanical description of Zwicky's "tired-light" theory.]

In the former case, the intensity of the photon flux of radiation is reduced as *gamma ray photons* are simply scattered out of a beam of *gamma* radiation. In the latter case, *photons* of longer wavelengths than those of *gamma rays* are seen as passing directly over the *free electrons* with therefore no change in a *photon*'s forward direction. Easily visualized as an example, a very long radio wavelength *photon* which passes over a *free electron* will certainly cause a radio frequency oscillation in the *electron*, while the radio wavelength *photon* continues along in its original straight path. In this long wavelength case, the intensity (*photon* flux) of a beam of *photons* remains unchanged. What is, however, expected is a minuscule reduction in each incident *photon's energy* (a *redshift*) as the *free electrons* are encountered, as will be demonstrated below by calculation. This mode of interaction is expected to hold not only for radio frequencies but over the entire frequency range of observed spectral lines all the way into the x-ray regime.

A further distinction is drawn here between these two separate cases. In high frequency *gamma ray Compton scattering*, *photons* can experience a change in their *frequency*, but a strong unshifted component remains. In contrast, *photons* of longer wavelengths than those of *gamma rays* as observed in the spectral lines from stars do not exhibit both frequency-shifted and unshifted light, rather *these photons are seen as all interacting with deep space free electrons in a wave-particle fashion*, where all *photons* are equally *redshifted* by the law of large numbers, each *photon* from a particular spectral line of a particular object encountering the same great number of *free electrons* in deep space.

As a note, in the situation of extremely dense radiation fields such as where very powerful lasers are used in laboratory *Thomson scattering* measurements of plasma densities, or in the extremely dense radiation fields in the interior of stars, a new interpretation of the manner is suggested here in which *photons* of longer *wavelengths* than those of *gamma rays* are actually interacting with *free electrons*. In these dense radiation fields, one has the impression that the powerful laboratory laser beam's *photons* can experience a change in their direction and be subtracted from the beam, as do *gamma ray photons* in *Compton scattering*. It is, however, suggested here that in these very dense radiation fields, the individual *photons* of longer *wavelengths* than those of *gamma rays* are not actually deflected from their straight-line paths, rather the *free electrons* are so massively agitated by radiation that a *free electron* reradiates *photons* at the same frequency of the radiation field only after taking a tiny bit of energy from each and every *photon* that passes directly over the *free electron*. The incident laboratory laser beam would then have no reduction in its *photon* flux but rather simply a tiny reduction in the *frequency* of each incident *photon*. The *cosmological redshift* might therefore be testable using laboratory lasers, although conditions in dense radiation fields in laboratory plasmas are radically different from those in deep space.

II. ANALYSIS OF THE REDSHIFT MECHANISM

The following discussion analyzes by calculation the (wave-particle) fashion in which *photons* of longer *wavelengths* than those of *gamma rays* may interact with deep space *free electrons*, resulting in the *redshifted* spectral lines of astronomical objects. A *photon* is an *electromagnetic wave* whose *electric field* component should cause an oscillation in any *free electron* over which passes the wave. The *electron* duly accelerated must *reradiate energy at the expense of the wave*. A *free electron* has been demonstrated by electromagnetic theory to have an effective area for reradiating the *energy* of an incident *electromagnetic wave*, the *Thomson scattering cross section*[7].

[7] Thomson, J. J. (1906). *Conduction of Electricity Through Gases.* Cambridge University Press, Cambridge, London.

Applying *Planck's relation* ($E = hf$) that the *energy* of a quantum of *electromagnetic energy* be proportional to its *frequency*, one expects then that the *frequency* of the *photon* is lowered in proportion to this reduction in its *energy*, i.e., a *redshift*.

[The *quantum mechanical* explanation!]

Much more massive are ions whose effect *is neglected*. From the point of view of electromagnetic wave theory, as a *photon* passes over a single *free electron* the *electron* is not displaced from its initial position but simply oscillates about its fixed position with the electric field of the wave, reradiating *energy* in a symmetric dipolar fashion, therefore not causing the wave to alter its forward direction. Note that in the calculation of the *Thomson scattering cross section*, the *electron* is taken as fixed in that the random velocities of *free electrons* are assumed small compared to the speed of light of the incident *photon*.

> [*Thomson scattering* is the elastic scattering of electromagnetic radiation by a free charged particle, as described by classical electromagnetism. It is the low-energy limit of *Compton scattering*: the particle's *kinetic energy* and *photon frequency* do not change as a result of the scattering. This limit is valid as long as the *photon energy* is much smaller than the *mass energy* of the particle: $v \ll mc^2/h$, or equivalently, if the *wavelength* of the light is much greater than the *Compton wavelength* of the particle (e.g., for *electrons*, longer *wavelengths* than hard x-rays).]

Also, in the calculation of the *Thomson scattering cross section*, the *electron* is taken as reacting only to the *photon*'s electric field component, the *electron*'s assumed *sub-relativistic* velocity allowing one to neglect the *magnetic* component of the *Lorentz force*.

> [The *Lorentz force* is a force exerted by an *electromagnetic field* on a *charged* particle.]

Any effect of the *electron*'s *dipole moment* is also neglected.

If now a *photon*, viewed as in itself an incident *electromagnetic wave*, traverses the rarefied deep space of the cosmos for billions of years, the *photon*'s wavefront slowly and eventually encounters vast numbers of *free electrons* one at a time, resulting in a cumulative *redshift* which can be calculated. The following equation expresses a fractional decrease (dI/I) in a plane *electromagnetic wave*'s energy flux I as the wave encounters an *electron* density n, where *electrons* have an effective cross-sectional area C, the well-known standard value for the *Thomson scattering cross section* of *electrons*. The equation follows immediately from the definition of the *Thomson scattering cross section*, which is the effective area of the *free electron* to reradiate energy from an incident *electromagnetic* wave (the *photon*) whose direction remains unchanged.

$$dI/I = -\,Cndx. \tag{1}$$

We can now integrate this equation, the integral of the right-hand side being the total cross-sectional area of all the *electrons* per m^2 of incident wave that the incident wave would be intercepting over a distance x. Completing the integration of both sides of the equation and then solving for I yields a standard exponential decay for I, the incident energy flux, over a distance x, where the exponential decay constant k has the value (1/Cn).

$$I \sim \exp(-x/k). \tag{2}$$

Assuming for illustrative purposes, an average *electron* density of 100 e/m^3 in *intergalactic space*, the exponential distance scale k (= 1/Cn) for the weakening (*redshifting*) *photon* calculates to be 16 x 10^9 light years. Dividing by the speed of light gives an exponential *redshifting* time scale of 16 x 10^9 years, *which is approximately the hypothetical "age of the universe" according to the Big Bang theory*. This provides a simple alternative explanation for the extremely *redshifted* edge of the visible universe, due to *wave-particle* scattering by *free electrons*, as opposed to the expansion hypothesis of the *Big Bang theory*.

Regarding the above assumption of an estimated 100 *free electrons*/m^3, note that any small uniform back ground of *free electrons* in deep space would have negligible effect on the observed dynamics of astronomical systems. Any higher or lower estimate for an average *free electron* density in deep space would, respectively, decrease or increase the above calculated *redshifting* distance scale. Estimated values for *Hubble's constant* have varied appreciably. By choosing a higher or lower figure for the average *free electron* density, one can precisely produce the same effect of any estimated value for *Hubble's constant* in that by either *redshifting* mechanism the linear distance versus *redshift* graphs for nearby measurable astronomical objects would coincide.

III. AGREEMENT WITH THE HUBBLE DIAGRAM

The precise coinciding of *redshift* graphs is quickly seen from the above exponential expression for the *redshifting photon*, where the *photon's frequency* f has the following dependence:

$$f \sim \exp(-Cnx). \tag{3}$$

For nearby astronomical objects, the *frequency* is therefore linear with distance.

$$f \sim (1 - Cnx). \tag{4}$$

Calculating *redshift*, we then arrive immediately at the following equation:

$$z = redshift = Cnx. \tag{5}$$

The Hubble expression for *redshift* is also linear with distance and precisely coincides with the above linear expression for *redshift* when simply equating the proportionality constants.

$$Cn = H/\text{speed of light}. \tag{6}$$

Taking one estimate of Hubble's constant to be the inverse of 13.7×10^9 years, the value of n is calculated to be $116 \ e/m^3$, the average *free electron* density that produces a linear distance versus *redshift* behavior which precisely coincides with that from Hubble's constant.

The above calculation demonstrates how just a small amount of *intergalactic free electrons* can result in the *cosmological redshift* observed in the spectral lines of astronomical objects.

Note that the emergent spectrum originates at the star's surface, and the *cosmological redshift* begins to increase thereafter, as the *photons* pass over vast numbers of *free electrons* in deep space after billions of years of travel. Further regarding the above assumption of approximately 100

157

e/m^3 in deep space, the average *electron* density in deep space has never been directly measured. The discovery of voids and supervoids in deep space make even more difficult the problem of directly measuring an effective average value for the density of *electrons* in deep space. The arguments presented in this present article are based on the implicit assumption that the *electron* density in deep space is homogeneous in space and time. Models of the *Big Bang theory* predict numbers for the *mass* density of the universe, but if one rejects the *Big Bang theory* and proposes alternative theories, the *Big Bang* based predictions for mass density are meaningless. The above determination of an average effective density of 116 e/m3 is supported by the above calculation which shows precise agreement with the current value for *Hubble's constant*.

As for laboratory confirmation of my above proposed explanation for the *cosmological redshift*, to detect a *redshift* in the laboratory would be difficult because *electron* densities normally attained in laboratory plasmas are far too low. However, the effect does appear to exist over astronomical distances where vast numbers of *free electrons* are available.

IV. CALCULATION OF THE DISTANCES TO THE FURTHEST OBSERVABLE OBJECTS

One now returns to the above predicted *redshifting frequency* dependence which was expressed by the following:

$$f \sim \exp(-Cnx). \tag{7}$$

From this *frequency* dependence, the *redshift* of the weakening *photon* is then immediately calculated yielding the following general formula (*redshift* = fractional change in *frequency*) which holds out to the furthest cosmological distances.

$$z = \text{redshift} = \exp(Cnx) - 1. \tag{8}$$

This *redshift* formula is therefore independent of the observed brightness of an astronomical object. It is also independent of the *photon*'s *frequency*, thereby admitting the same *redshift* measurement in any *wavelength* band of observed spectral lines. Linear at nearby distances in precise agreement with Hubble's data, as shown above, we now find that at great distances the *redshift* should increase exponentially.

Solving this expression for *redshift* one finds the following general equation for determining the *distance* x to cosmological objects based on their *redshifts*.

$$x = (1/Cn) \ln (\text{redshift} + 1). \tag{9}$$

At great distances, using the above calculated $n = 116$ *free electrons*/m^3 in deep space, one sees that for a *redshift* of 1.72, the distance of an astronomical object reduces to the following:

$$x = 1/Cn = (\text{speed of light})/H = 13.7 \ 10^9 \text{ light years. (10)}$$

Type 1a supernovae with *redshift* of 1.72 should then be at a distance of 13.7×10^9 light years. Distant *Type 1a supernovae* are observed to be much dimmer than their *redshifts* would normally indicate, leading one to believe them to be further than 13.7×10^9 light years. However, the cumulative effect of cosmic dust at great distances is presumed to be responsible for their dimness and for their divergence from the linear *Hubble relation* at high *redshifts*[8,9].

[8] Kirshner, R. P. (1999). *Proc. Natl. Acad. Sci.,* 96, 4224.
[9] Aguirre, A. (1999). *Astrophys. J.,* 525, 583.

Using distance modulus to calculate the distance of nearby astronomical objects is reliable, but at great distances absorption coefficients of cosmic dust make distance modulus measurements uncertain. The above exponential expression for *redshift* allows distance calculations for the furthest observed astronomical objects without needing any corrections for cosmic dust. The arrival of a single *photon* from a particular spectral line in principle allows the calculation of the distance to the furthest observable object. It matters not how many *photons* arrive, the above *redshift* expressions being independent of observed *brightness*. The above exponential expression for *redshift* circumvents the problem of dust extinction when measuring the furthest cosmological distances.

Observations of *time dilation* in *supernova* light curves are here regarded as inconclusive, such studies perhaps involving systematic errors in their interpretation or treatment of data, possibly in their sampling of intrinsically brighter *supernovae* at high *redshifts* while ignoring these dimming effects of cosmic dust. The surface brightness test is also regarded here as inconclusive in view of these heretofore neglected effects of cosmic dust at high *redshifts*.

V. FINAL COMMENTS

A new theoretical model has been presented here which accounts for the *cosmological redshift* in a static universe.

[This is a reformulation of Zwicky's *"tired-light" theory*.]

This new theoretical *redshift* model is simpler than the hypothesis of expanding space as derived from the [*relativistic*] gravitational *field equation*. This new explanation for the *cosmological redshift* also provides a solution to Olbers' paradox, a *photon* slowly *redshifting* to *frequencies* not capable of stimulating the human eye. This new *non-Doppler* explanation for the *cosmological redshift* also permits for the first time distance measurements to the furthest observable astronomical objects. Not only do these newly allowed distance measurements circumvent the problem of cosmic dust, they also are no longer subject to the question of the *Big Bang*'s adjustable scale factors. Without the *Big Bang theory* comes a new postmodern cosmology where the universe is seen as presumably infinite spatially and temporally, which necessarily implies a new dynamic equilibrium cosmology.

It is suggested here that research is directed into identifying the processes which maintain this equilibrium, namely, processes whereby entropy must be recycled and starlight returns to matter, both these conditions possibly satisfied by deep space *pair production* processes. The *cosmic microwave background* should be reconsidered as due to the temperature of space as first calculated in 1896 by Nobel Prize winner Charles Édouard Guillaume[10].

[10] Guillaume, C. -E. (1896). *La Nature*, 24, 234.

Without the *Big Bang theory*, a symmetric universe with equal amounts of *matter* and *antimatter* can now be considered, evidenced in the *cosmic gamma ray* background radiation. A picture then emerges of *matter-antimatter annihilation* keeping an eternal universe churning, all matter unable to coalesce to any particular point. This then immediately offers a new direction of research into *gamma ray* bursts, *quasars*, *blazars*, and other extremely energetic objects, possibly explainable by various scenarios of *matter-antimatter annihilation*. This would avoid having to use the unphysical mathematical singularities inherent in the *gravitational field equation*, on which have been based models of astronomical *black holes* as well as the initial hypothetical *Big Bang* itself.

The effect of a strong magnetic field on a dielectric.

Mamas, D. L. (2010) is reminiscent of a recent explanation by the author of the *Faraday effect*, in which, when addressed in terms of *non-relativistic quantum electrodynamics* (or New Physics) "the *electrons* in the *dielectric* under the influence of the strong *magnetic field* line up according to their *spin*. The motion thus effected will be *circular*; and rotating *electric charges* create a *magnetic field*. So, the circularly moving *electrons* will create their own *magnetic field* in addition to the external *magnetic field* on them. There will thus be *two different cases*: the created field will be parallel to the external field for one *circular polarization*, and in the opposing direction for the other polarization direction – thus the net *magnetic field* is enhanced in one direction and diminished in the opposite direction by the same amount. In a *linearly polarized* magnetic field this will result in the rotation of the polarization depending on the propensity for the dielectric to create the magnetic field, ultimately on the *dielectric constant* of the dielectric.".

However, although the consequences are similar, unlike in Mamas, D. L. (2010), this is a *quantum mechanical* description, rather than a pseudo classical explanation based on a *"wave-particle* model" in which the *free electrons*, over which the *electromagnetic wave* passes over, *oscillate and reradiate energy in accordance with Maxwell's equations* at the expense of the *photon* which can then be expected to *redshift*. In the Mamas theory it is the effect on the *electromagnetic wave*, rather than the *magnetic field*, of the resulting oscillation of free electrons and reradiation according to Maxwell's equations which creates the change in the electromagnetic wave and consequent *redshift*.

Described in terms of *quantum mechanics*, the beam of *circularly polarized photons* rotating in one direction corresponds to one quantum *spin state* of the *photon*. The beam of *circularly polarized photons* rotating in the other direction corresponds to the other quantum *spin state*. In quantum theoretical terms, when a light beam is *circularly polarized*, each of its *photons* carries a *spin angular momentum* of $\pm \hbar$,

162

where $\hbar$ is the reduced Planck constant and the ± sign is positive for left and negative for right *circular polarizations* (adopting the convention from the point of view of the receiver most commonly used in optics). This *spin angular momentum* is directed along the beam axis (parallel if positive, antiparallel if negative).

The *electrons* in the *dielectric* under the influence of the strong *magnetic field* line up according to their *spin*. The motion thus effected will be *circular*; and rotating *electric charges* create a *magnetic field*. So, the circularly moving *electrons* will create their own *magnetic field* in addition to the external *magnetic field* on them. There will thus be *two different cases*: the created field will be parallel to the external field for one *circular polarization*, and in the opposing direction for the other polarization direction – thus the net *magnetic field* is enhanced in one direction and diminished in the opposite direction by the same amount. In a *linearly polarized* magnetic field this will result in the rotation of the polarization.

Underwood, T. G. (October 12, 2024). The Faraday effect.

Unpublished.

The Faraday effect was first mentioned by Michael Faraday in his daily notebook on September 13, 1845, paragraph #7504:

> "... , when the contrary magnetic poles were on the same side, there was an effect produced on the polarized ray, and thus magnetic force and light were proved to have relation to each other. ...",

of which he sent a detailed account to the Royal Institution on October 29, 1845[1].

> [1] Received November 6, 1845; read November 20, 1845; published in Experimental Researches in Electricity. Nineteenth Series. *Phil. Trans.*, I, §26, 1-20, 2146-2242. Faraday, M. (1845). "On the magnetization of light and the illumination of magnetic lines of force. i. Action of magnets on light. ii. Action of electric currents on light. iii. General considerations".

In this account, Faraday, noted the following:

(1) # 2154: "The character of the force thus impressed upon the diamagnetic is that of rotation ...";

(2) # 2155: ... "The direction was always the same for the same line of magnetic force;

(3) # 2160: "... They give the diamagnetic the power of rotating the ray; and the law of this action on light is, that if a magnetic line of force be going from a north pole, or coming from a south pole, along the path of a polarized ray coming to the observer, it will rotate that ray to the right-hand; or, that if such a line of force be coming from a north pole, or going from a south pole, it will rotate such a ray to the left-hand.";

(4) # 2199: "*... When an electric current passes round a ray of polarized light in a plane perpendicular to the ray, it causes the ray to revolve on its axis, as long as it is under the influence of the current, in the same direction as that in which the current is passing*;

(5) #2222: "The relation existing *between polarized light and magnetism and electricity*, is even more interesting than if it had been shown to exist with common light only. It cannot but extend to common light; ...";

(6) # 2223: "... it is the *magnetic lines of force* only which are effectual on the rays of light, and they only (in appearance) when parallel to the ray of light, or as they tend to parallelism with it.";

(7) #2224: "The magnetic forces do not act on the ray of light directly and without the intervention of matter, but through the mediation of the substance in which they and the ray have a simultaneous existence; the substances and the forces giving to and receiving from each other the power of acting on the light. This is shown by the non-action of a vacuum, of air or gases; and it is also further shown by the special degree in which different matter's possess the property. ...";

(8) #2225: "2225. Recognizing or perceiving *matter* only by its powers, and *knowing nothing of any imaginary nucleus,* abstract from the idea of these powers, the phenomena described in this paper much strengthen my inclination to trust in the views I have on a former occasion advanced in reference to its nature.";

(9) #2226: "It cannot be doubted that *the magnetic forces act upon and affect the internal constitution of the diamagnetic*, just as freely in the dark as when a ray of light is passing through it; though the phenomena produced by light seem, as yet, to present the only means of observing this constitution and the change. Further, any such change as this must belong to opaque bodies, such as wood, stone, and metal; for as *diamagnetics*, there is no distinction between them and those which are transparent. ...";

(10) #2227: "... the molecular condition of these bodies, when in the state described, must be specifically distinct from that of magnetized iron, or other such matter, and must be *a new magnetic condition*; and as the condition is a state of tension (manifested by its instantaneous return to the normal state when the magnetic induction is removed), so the *force* which the matter in this state possesses and its mode of action, must be to us a *new magnetic force or mode of action of matter*.";

(11): "Perhaps this state is a state of *electric tension tending to a current*; as in magnets, according to Ampere's theory, the state is a state of *current*. ...";

(12): #2231: "... Oil of turpentine will rotate a ray of light, the power depending upon its particles and not upon the arrangement of the mass. *Whichever way a ray of polarized light passes through this fluid, it is rotated in the same manner*; and rays passing in every possible direction through it simultaneously are all rotated with equal force and according to one common law of direction; *i.e. either all right-handed or else all to the left. Not so with the rotation superinduced on the same oil of turpentine by the magnetic or electric forces: it exists only in one direction, i.e. in a plane perpendicular to the magnetic line*; and being limited to this plane, it can be changed in direction by a reversal of the direction of the inducing force. The direction of the rotation produced by the natural state is connected invariably with the direction of the ray of light; but the power to produce it appears to be possessed in every direction and at all times by the particles of the fluid: *the direction of the rotation produced by the induced condition is connected invariably with the direction of the magnetic line or the electric current*, ...".

In view of the lack of theoretical knowledge about the composition of matter at that time in 1845, this was an extraordinary demonstration of experimental physics at its best.

Today, the Faraday effect is generally understood to be caused by left and right *circularly polarized* waves propagating at slightly different speeds, a property known as *circular birefringence*. Since a *linear polarization* can be decomposed into the *superposition* of two equal-amplitude *circularly polarized* components of opposite handedness and different phase, the effect of a relative phase shift, induced by the Faraday effect, is to rotate the orientation of a wave's *linear polarization.*

The phenomenon of *polarization* arises as a consequence of the fact that light behaves as a two-dimensional *transverse wave. Electromagnetic waves* are *transverse waves* without requiring a medium. A *transverse wave* is a wave that oscillates perpendicularly to the direction of the wave's advance. All waves move energy from place to place without transporting the matter in the transmission medium if there is one. *Circular polarization* of an *electromagnetic wave* is a polarization state in which, at each point, the *electromagnetic field* of the wave has a constant magnitude and is rotating at a constant rate in a plane perpendicular to the direction of the wave. *Circular polarization* occurs when the two orthogonal electric field component vectors are of equal magnitude and are out of phase by exactly 90°, or one-quarter wavelength.

The strength and direction of an *electric field* is defined by its *electric field vector*. In the case of a *circularly polarized wave*, the tip of the *electric field vector*, at a given point in space, relates to the *phase* of the light as it travels through time and space. At any instant of time, the *electric field vector* of the wave indicates a point on a helix oriented along the direction of propagation. A *circularly polarized wave* can rotate in one of two possible senses: right-handed circular polarization in which the *electric field vector* rotates in a right-hand sense with respect to the direction of propagation, and left-handed circular polarization in which the vector rotates in a left-hand sense.

An *electromagnetic wave* is said to have *circular polarization* when its *electric* and *magnetic fields* rotate continuously around the beam axis during propagation. The *circular polarization* is left (L) or right (R) depending on the field rotation direction and, according to the convention used: either from the point of view of the source, or the receiver.

167

In *circularly polarized* light the direction of the *electric field* rotates at the *frequency* of the light, either clockwise or counter-clockwise. In a *dielectric material*, this *electric field* causes a force on the *charged* particles that compose the material. Because of their large *charge to mass ratio*, the *electrons* are most heavily affected. The motion thus effected will be *circular*, and circularly moving *charges* will create their own *magnetic field*. This additional *magnetic field* will be in the opposite direction to the applied external *magnetic field*.

Dielectric materials are composed of polar molecules. Such molecules have spatially separated positive and negative *electric charge*. The molecules may be either bound in one position (solids) or free to move (liquids and gases). When a *dielectric* material is placed in an *electric field*, *electric charges* do not flow through the material as they do in an electrical conductor, because they have no loosely bound, or free electrons that may drift through the material, but instead they shift, only slightly, from their average equilibrium positions, causing *dielectric polarization*. Applied *electric fields* align the molecules. The resulting *charge* displacement reduces the *electric field* in the material and modifies fields in the vicinity of the *dielectric*. There are corresponding magnetic field effects in *paramagnetic* and *ferromagnetic* materials.

In *circularly polarized* light the direction of the *electric field* rotates at the *frequency* of the light, either clockwise or counter-clockwise. In a *dielectric material*, this *electric field* causes a force on the *charged* particles that compose the material. Because of their large *charge to mass ratio*, the *electrons* are most heavily affected.

Because of *dielectric polarization*, positive charges are displaced in the direction of the field and negative charges shift in the direction opposite to the field. This creates an internal *electric field* that reduces the overall field within the *dielectric* itself. If a dielectric is composed of weakly bonded molecules, those molecules not only become *polarized*, but also reorient so that their symmetry axes align to the field.

The three *electric-field* components of left and right *circularly polarized* plane waves propagating in the z direction in complex notation are:

$$|L> = 1/\sqrt{2} \begin{pmatrix} 1 \\ i \\ 0 \end{pmatrix} e^{i(kz-\omega t)}$$

$$|R> = 1/\sqrt{2} \begin{pmatrix} 1 \\ -i \\ 0 \end{pmatrix} e^{i(kz-\omega t)}$$

where k is the *wave number* (cycles per unit distance) and ω the *angular frequency*.

The *linear polarized light* that is seen to rotate in the *Faraday effect* can be seen as consisting of the *superposition* of a right- and a left- *circularly polarized* beam. We can look at the effects of each component (right- or left-polarized) separately, and see what effect this has on the result.

Linear polarization or *plane polarization* of *electromagnetic radiation* is a confinement of the *electric field vector* or *magnetic field vector* to a given plane along the direction of propagation. The orientation of a *linearly polarized electromagnetic wave* is defined by the direction of the *electric field vector*. If the *electric field vector* is vertical (alternatively up and down as the wave travels) the radiation is said to be *vertically polarized*.

There will thus be *two different cases*: the created *magnetic field* will be parallel to the external field for one *circular polarization,* and in the opposing direction for the other polarization direction – thus the net *magnetic field* is enhanced in one direction and diminished in the opposite direction. This changes the dynamics of the interaction for each beam and one of the beams will be slowed more than the other, causing a *phase difference* between the left- and right-polarized beam. When the two beams are added after this *phase shift,* the result is again a *linearly polarized* beam, but with a *rotation* of the *polarization vector.*

The direction of polarization rotation depends on the properties of the material through which the light is shone. A full treatment would have to take into account the effect of the external and radiation-induced fields on the *wave function* of the *electrons*, and then calculate the effect of this change on the *refractive index* of the material for each *polarization*, to see whether the right- or left-circular polarization is slowed more.

The *classical calculation* of the resulting *wave numbers* for right- and left-handed *circularly polarized* waves of the same frequency gives slightly different *wave numbers*,

$$k_\pm \simeq k \pm \Delta,$$

where $k = \omega[1 - (1/2)\, \omega_p^2/\omega^2]/c$, $\Delta k = (1/2)\,(\omega_p^2/\omega^2)\Omega/c$, and Ω is the *gyration frequency* of free *electrons* in the *magnetic field*. [See https://farside.ph.utexas.edu/teaching/em/lectures/node101.html.]

In *classical theory*, it is this peculiarity of the wave formulation that results in the *Faraday rotation* of *linearly polarized* waves. (Faraday would have been amazed.)

However, this explanation is *classical physics* applied to the wave nature of light. It does not take fully into account the *quantization* of light.

Described in terms of *quantum mechanics*, there is a slightly different result. The beam of *circularly polarized photons* rotating in one direction corresponds to one quantum *spin state* of the *photon*. The beam of *circularly polarized photons* rotating in the other direction corresponds to the other quantum *spin state*. In quantum theoretical terms, when a light beam is *circularly polarized*, each of its *photons* carries a *spin angular momentum* of $\pm\,\hbar$, where $\hbar$ is the reduced Planck constant and the $\pm$ sign is positive for left and negative for right *circular polarizations* (this is adopting the convention from the point of view of the receiver most commonly used in optics). This *spin angular momentum* is directed along the beam axis (parallel if positive, antiparallel if negative).

The *spin angular momentum* of light is the component of angular momentum of light that is associated with the *quantum spin* and the *rotation* between the polarization degrees of freedom of the *photon*.

Spin is the fundamental property that distinguishes the two types of *elementary particles*: *fermions*, with half-integer *spins*; and *bosons*, with integer *spins*. *Photons*, which are the *quanta* of light, have been long recognized as spin-1 *gauge bosons*. The *polarization* of the light is commonly accepted as its "intrinsic" *spin* degree of freedom. However, in free space, only two *transverse polarizations* are allowed. Thus, the *photon spin* is always only connected to the two *circular polarizations*. To construct the full *quantum spin operator* of light, *longitudinal polarized photon* modes have to be introduced.

The *electrons* in the *dielectric* under the influence of the strong *magnetic field* line up according to their *spin*. The motion thus effected will be *circular*; and rotating *electric charges* create a *magnetic field*. So, the circularly moving *electrons* will create their own *magnetic field* in addition to the external *magnetic field* on them. There will thus be *two different cases*: the created field will be parallel to the external field for one *circular polarization*, and in the opposing direction for the other polarization direction – thus the net *magnetic field* is enhanced in one direction and diminished in the opposite direction by the same amount. In a *linearly polarized* magnetic field this will result in the rotation of the polarization depending on the propensity for the dielectric to create the magnetic field, ultimately on the *dielectric constant* of the dielectric.

Then there is the issue of whether *quantum entanglement* of the quantum *spin states* of either *photons* or the *electrons* in the *dielectric* has any effect, as it does in the *valance bond* and in *ferromagnetism*.

Shaoa, M-H. (April, 2013). The energy loss of photons and cosmological redshift*.

Published online 8 April 8, 2013; http://dx.doi.org/10.4006/0836-1398-26.2.183.

Xinjiang Astronomical Observatory (XAO), Chinese Academy of Sciences (CAS), Key Laboratory of Radio Astronomy, 150, Science 1-Street, Urumqi, Xinjiang 830011, People's Republic of China.

* This work is partially supported by the Key Laboratory of Radio Astronomy, CAS.

Received: March 13, 2012.
Accepted: February 4, 2013.

This is another rather bizarre article in which Zwicky's *"tired-light"* theory is presented in terms of a simplistic interpretation of the wave-particle nature of the *photon* and classical electromagnetic theory. This is rather surprising as it was published by Xinjiang Astronomical Observatory, Key Laboratory of Radio Astronomy, Chinese Academy of Sciences, though not by a mainstream theoretical physics institution. It is included for the same reasons as Mamas, D. L. (2010).

Abstract: The *"tired-light" hypothesis* interprets the *cosmological redshift* on an assumption of *energy loss* of *photons* interacting with massive particles. *A new mechanistic model of "tired-light" is proposed by establishing a differential equation through the analysis of a photon's energy loss based on the classical electromagnetic theory and principles of the mass–energy equivalence and the wave–particle duality.* The *photon* transfers some *energy* to massive particles by *Lorentz electric force* in their interaction. The expanding result of the equation's solution is a formula for *redshift*, which shows a relationship of *redshift* to the *wavelength* of light and the *number* of massive particles that the *photon* encountered. The model agrees with the observational data in the cited literature by the relationship of *redshift* to *wavelength* of light. And, the relationship of *redshift* to the *number* of massive particles encountered by the *photons* explains the *"limb effect"* on the *solar disc*, the signal *redshift* of Pioneer 6, and the large *redshifts* of *quasars*, all of which cannot be explained by the *Doppler effect*. *The new model provides a basis for the hypothesis of "tired-light" to become a reasonable theory.*

I. INTRODUCTION

The relationship of *cosmological redshift* (CR) to the *distance* traveled by the light was discovered more than 80 years ago by Hubble[1].

[1] Hubble, E. (1929). *Proc. Natl. Acad. Sci.*, 15, 168.

The *Doppler effect* was accepted to interpret the *cosmological redshift* (CR) after it had been employed to study the movement of double stars, that of the Sun in our galaxy, and the rotation of our galaxy. In the three cases there are blueshift and redshift together, and the magnitude of the shifts is independent of the distance to the source. But, contrarily, in the case of CR there is only *redshift* but no *blueshift*, and the magnitude of the *redshift* of CR is *proportional to the distance* (see Ref. 2). Reber[2] noted that "Clearly the interpretation of these spectral shifts as representing relative motion was dubious".

[2] Reber, G. (1977). Endless Boundless Stable Universe (University of Tasmania, Hobart).

More than that, the *Doppler effect* cannot explain some phenomena. The first is the *"limb effect"*, i.e., the variation of *redshift* from the center to the limb of the *solar disc* discovered more than a century ago (see Ref. 3).

[3] LoPresto, J. C., Schrader, C., & Pierce, A. K. (1991). *Astrophys. J.*, 376, 757.

The second is the signal *redshift* of Pioneer 6. When Pioneer 6 on its orbit at the other side of the Sun approaching the limb of *solar disc* in

November 1968, the signal from it to the observer on the Earth gave an additional frequency shift or redshifted. The third important phenomenon that cannot be explained by the *Doppler effect* is the large *redshift* of *quasars*.

Zwicky,[4] about half a year after Hubble's discovery, proposed the *"tired-light" hypothesis* as an alternative interpretation of *cosmological redshift* (CR) to the *Doppler effect*.

[4] Zwicky, F. (1929). *Proc. Natl. Acad. Sci.*, 15, 773, see above.

He assumed a force due to *gravitational* "drag" of *masses* in space acted on light passing nearby. Thus, the light would transfer *momentum* and *energy* to the masses, and the *frequency* of the light changed in the process. This idea was vaguely related to known physical principles. So, the search for a mechanism of *"tired-light"* has never ceased since then. As early as in 1962, de Broglie expressed his thoughts about *"tired-light"* in that the *redshift* was due to attenuation of *photons* when they interacted with *absorbing matter*[5].

[5] Assis, A. K. T. (1993). *Progress in New Cosmologies: Beyond the Big Bang*, edited by H. C. Arp, C. R. Keys, and K. Rudncki, Plenum Press, New York, p. 153.

Various mechanisms were subsequently proposed and discussed[6–11].

[6] Pecker, J. C. & Vigier, J. P. (1987). *Observational Cosmology*, edited by A. Hewitt,
G. Burbidge, and L. Z. Fang, (IAU, Beijing), p. 507.
[7] Marmet, P. (1988). *Phys. Essays*, 1, 24.
[8] Ghosh, A. (1991). *Apeiron*, 9–10, 95.
[9] Assis, A. K. T. (1992). *Apeiron*, 12, 13.
[10] Rabounski, D. (2009). *Prog. Phys.* 1, L1.
[11] Sorrell, W. H. (2009). *Astrophys. Space Sci.*, 323, 205.

But none of the models were directly based on known physical principles. Some of them have inherent problems such as the deflection of *photons* by *Compton scattering*[7]. Authors in Refs. 5, 6, 10, and 11 gave CR by "tired-light" in the form $Z = \exp(H_0 d/c) - 1$. Most recently, Mamas[12] predicted a new mechanism of interaction of *photons* with free electrons in deep space, for the *"tired-light" hypothesis*.

[12] Mamas, D. L. (2010). An explanation for the cosmological redshift. *Phys. Essays*, 23, 326, see above.

In what follows, the interaction process between a *photon* and massive particles with energy transfer will be analyzed to develop a new model of *"tired-light"*.

II. ENERGY LOSS AND THE REDSHIFT OF PHOTONS

A. *The densities of photons and massive particles*

The *intergalactic space* is not empty[13].

[13] Mo, H., Bosch, F, & White, S. (2010). *Galaxy Formation and Evolution*. Cambridge University Press, Cambridge, p. 689.

There are many kinds of massive particles drifting in the intergalactic space (*intergalactic matter*, or IGM), more massive particles in the interstellar space *(interstellar matter*, or ISM), and many more massive particles around stars. *A photon is a section of an electromagnetic wave* [??? !] It must meet some massive particles while traveling through space. The *photon* is expected to interact with the massive particles, accompanied by *energy* transfer. I consider here the most likely process, where energy is transferred from the *photon* to the massive particles.

Saalmann and Rost[14] demonstrated the *absorption of energy* from a laser field into a xenon cluster with full dynamical microscopic calculations.

[14] Saalmann, U. & Rost, J. M. (2003). *Phys. Rev. Lett.*, 91, 223401.

Kim *et al.*[15] had observed spectral *redshifts* of a laser beam interacting with the gas jet of an argon cluster.

[15] Kim, K. Y., Alexeev, I., Antonsen, T. M., Gupta, A., Kumarappan, V., & Milchberg, H. M. (2005). *Phys. Rev.*, A 71, 11201.

The xenon and argon clusters are composed of atoms, i.e., massive particles, and the laser is a kind of *photons* of narrow frequency band. Mamas[12] discussed the interaction mechanism between particles and *photons* with a new viewpoint that *the interaction is in a wave–particle fashion.* In a simplified case, Mamas[12] neglected particles more massive than *electrons* and analyzed the interaction of *photons* and *free electrons* in deep space. He reached a conclusion that the energy transfer from *photons* to free electrons would induce a *redshift* of the *photons*.

All the studies of Saalmann and Rost,[14] Kim et al.,[15] and Mamas[12] have shown that there is *energy transfer* in the *interaction* process of *photons* with massive particles. It is realized here that all massive particles, including atoms and molecules, are possible to be affected by *photons* in deep space.

The *energy* of a *photon* is determined by the Planck constant and the *frequency* of the light, i.e., $E = h\nu$. It can also be expressed as $E = hc/\lambda$, where c is the *velocity* of light, which is considered as a constant here. A *photon* and a massive particle (i.e., an atom or a molecule) should be treated equally from two basic viewpoints of the *mass–energy equivalence* and the *wave–particle duality*. *In this sense, a photon has an equivalent mass, and both a photon and a massive particle are physical objects but with different densities.* In the following, I will calculate the *densities* of a *photon* and massive particles.

A photon is *a section of an electromagnetic wave* with a certain *wavelength* k. [??? !] So, for the convenience of the analysis below, *I imagine a photon in the shape of a ball*, the simplest shape in a three-dimensional space, of diameter k. It possesses the shape of a particle and the characteristic of wave.

From the formulas $E = Mc^2$, and $E = hc/\lambda$, the mass of a *photon* is

$M = h/\lambda c$. *The volume of a photon is* $\pi\lambda^3/6$. [???] *The wavelength* of visible light is about 4 to 7 x 10^{-7} m. *Let the median wavelength* λ = *5.5 x 10^{-7} m represent the size of a photon.* The *Van der Waals radius* of a hydrogen atom is 1.2 x 10^{-10} m, so its size is 2.4 x 10^{-10} m. And the *mass* of a hydrogen atom is 1.67 x 10^{-27} kg. So, with h = 6.6 x 10^{-34} kgm^2 s^{-1} and c = 3 x 10^8 m s^{-1}, and *assuming a molecule is 100 times the mass of a hydrogen atom and 10 times the size of a hydrogen atom in all dimensions,* I obtain the densities

D_{photon} = 4.6 x 10^{-17} kg m^{-3};
$D_{hydrogen}$ = 231 kg m^{-3};
$D_{molecule}$ = 23 kg m^{-3}.

I take the mean value of $D_{hydrogen}$ and $D_{molecule}$ as the *density* of a general massive particle, then D_{mass} = 1.3 x 10^2 kg m^3, and I have

D_{photon}/D_{mass} = 3.5 x 10^{-19}.

Here, we should note that *a photon of visible light is about 200–2,000 times larger than a massive particle.*

B. *Interaction mechanism of photons and massive particles*

When a *photon* meets a massive particle, i.e., an atom or a molecule, *the massive particle will penetrate the photon, since the photon's density is 3.5 x 10^{-19} times lower and its size is 200–2,000 times larger than those of the massive particle.* [???] We know that, by both practical experience and logical reasoning, when two physical objects with different densities meet each other with enough relative velocity, generally speaking, the one smaller in size and of higher density will penetrate the other. *The massive particle penetrates through the photon* [???] but does not affect the traveling direction of the *photon* so that the traveling direction of the *photon* is unchanged after interacting with massive particles. *This is just the manner of photons of visible light interacting with the massive particles of air on the Earth.* [???]

Some massive particles may be *charged*. The electric neutral ones may have instantaneous polarity. Both of them should be affected by an *electromagnetic field. A photon is a section of an electromagnetic wave.* [???] Then, the massive particles will be affected by the electromagnetic field of *photons.* The *Lorentz force* that an *electromagnetic field* acts on a *charged* or *polarized* particle is composed of the *electric force* and the *magnetic force.* While a massive particle *penetrates a photon,* the *photon* will vibrate the massive particle by the Lorentz force and pass some energy to the massive particle by the *electric force.* [???] After the interaction, a tiny part of the energy of the photon should be transferred to the massive particle. *The photon's energy decreased from $h\nu_0$ to $h\nu$, and the size of the photon increased from λ_0 to λ.* Thus, *the photon got redshifted.* [???]

C. *Decrease of a photon's energy*

A *photon* will transfer more *energy* to a massive particle in a longer interaction time. *A larger photon has a longer interaction time with a massive particle, so it transfers a larger part of its energy to the massive particle than a smaller photon does. So, the energy loss of a photon is proportional to its size.* [???] After a *photon* has been emitted, it meets a number of massive particles on its way to an observer. The greater the number of massive particles it meets, the more *energy* it loses. So, the *photon* loses energy proportionally to the number of massive particles it meets.

When a *photon* of size λ and energy E meets a massive particle, *the massive particle penetrates through the photon ball in a time $t = \lambda/c$,* where c is the speed of light. [???] Meanwhile, the *photon* transfers a tiny amount of energy $\delta(E)$ to the massive particle. I define a coefficient k = $\delta(E)/E\lambda$, equal to the rate of *energy loss* of the *photon* per *unit length.* Here, the coefficient k *is conceptually designed.* Through theoretical or experimental approaches, it may accept a value for a further study.

If a *photon,* of size λ_0 and *energy* E when emitted, meets N massive particles and transfers a part of its *energy* to the massive particles on its way, supposing all kinds of massive particles interact equally with the *photon*, I obtain a differential equation for the *energy* of the *photon* with coefficient k as follows:

$$dE/dN = - k\lambda_0 E. \tag{1}$$

The solution of this equation is

$$E = E_0/\exp(kN\lambda_0) \tag{2}$$

from the condition $E = E_0$ when $N = 0$, where E_0 is the original *energy* of the *photon* when emitted. The *energy loss* of the *photon* is $\Delta E = E_0 - E$. So, I obtain

$$\Delta E = E_0 (1 - 1/\exp(kN\lambda_0)). \tag{3}$$

The expression for the *redshift* is $Z = (\lambda - \lambda_0)/\lambda_0$, and it can be written as $Z = h(v - v_0)/v_0 = (E_0 - E)/E$. Then, I have

$$Z = \Delta E/E. \tag{4}$$

Since $\Delta E/E = (E_0 - E)/E = E_0/E - 1$, from Eqs. (2) and (4), I get

$$Z = \exp(kN\lambda_0) - 1. \tag{5}$$

From Eq. (5), we see that the *redshift* is related to N, the number of massive particles that the *photon* has met, and λ_0, *the original size of the photon* [???] Equation (4) shows that *the redshift expresses the degree of energy loss of the photon*; for example, $Z = 1$ means that $\Delta E = E$, i.e., *the photon has lost half of its original energy.*

D. *The gravitational redshift*

Let us call the *redshift* induced by the process of *energy loss of photons* described in Section II C as *tired-light redshift* (which is referred to as "TR" here). In addition to the TR, the *gravitational redshift* (GR) must be considered in the analysis and evaluation of *cosmological redshift* (CR). The GR is different from the TR in nature. It has no relation to the interaction of *photons* with massive particles. So, Eq. (5) should have the form

$$Z = \exp(kN\lambda_0 + \lambda_g) - 1. \tag{6}$$

where λ_g denotes the part of the *wavelength* change induced by the *gravitational effect*. (We know that the GR equals GM/c^2R. Here, it is also expressed as λ_g/λ_0.) With $u = \lambda_g$, Eq. (6) becomes

$$Z = \exp(kN\lambda_0 + u) - 1. \tag{7}$$

Now, we have two expressions for the *cosmological redshift* (CR). Equation (5), comparatively simple, considers only the effect of *tired-light redshift* (TR), whereas Eq. (7) considers both the effects of *tired-light redshift* (TR) and *gravitational redshift* (GR).

III. FEATURES OF THE MODEL AND EVIDENCE

A. *Redshift versus wavelength*

A *photon* emitted from a star undergoes a continuous process of *energy loss* on its journey by interacting with massive particles before it reaches an observer on the Earth. It encounters massive particles, in the corresponding spaces, of (1) the atmosphere around the star, (2) the *interstellar medium* (ISM) in the galaxy to which the star belongs, (3) the *intergalactic medium* (IGM), (4) the *intergalactic medium* (ISM) of our galaxy, and (5) the *atmosphere* of the Earth. In the five parts, the *intergalactic medium* (IGM) is the main one in which the *photon* meets massive particles. Although the massive particles are sparsely distributed in the *intergalactic space*, the space is so vast to compare with the other spaces. Therefore, *the cosmological redshift (CR) is mainly induced by the interaction of photons with massive particles of the intergalactic medium* (IGM).

Since the massive particles of the *intergalactic medium* (IGM) are sparsely distributed, it may be supposed that the *photons* of different *wavelengths* emitted from a certain source meet the same *number* of massive particles in the *intergalactic space* along the line of sight of an observer on the Earth. Further, I suppose the *photons* meet roughly a same *number* of massive particles on their entire journey from being emitted to being observed. So, the *number* of massive particles N that the *photons* meet can be considered as a constant. Thus, N can be included in the coefficient k, and Eq. (7) takes the form

$$Z = \exp(k\lambda_0 + u) - 1. \tag{8}$$

The first-order approximation to Eq. (8) is

$$Z = k\lambda_0 + u. \tag{9}$$

The Eqs. (7) – (9) have shown that the *redshift* is a function of the *wavelength* of light, for which some evidences are provided in the following.

1. *Early evidence for the new model*

Zwicky[4] noticed the phenomenon of a relation between *redshift* and *wavelength*. He remarked, "Some exceptions have been found, suggesting that $\Delta v/v$ for H_β is somewhat greater than for H_γ". Here, $\Delta v/v = \Delta\lambda_0/\lambda_0$ is the *redshift*, and the value of λ_0 for H_β is greater than that for H_γ. What Zwicky noticed is just the relation of Eq. (9), i.e., a larger *redshift* is related to a longer *wavelength*.

Wilson[16] reported an observational result for the Seyfert galaxy NGC4151.

[16] Wilson, O. C. (1949). *Publ. Astron. Soc. Pac*, 61, 132.

He noticed a "slight apparent trend of *velocity* with *wavelength*". (The *redshift* is usually expressed as a *recession velocity* when interpreted in terms of the *Doppler effect*.) However, he obtained a *mean radial velocity* of 967 kms^{-1}.

…

Although the *cosmological redshift* (CR) of light from a source is considered to be a certain value by the Doppler interpretation, *redshift* differences of lines have been known and studied for a long time and some explanations were given[4,16–20].

[17] Espey, B. R., Carswell, R. F., Bailey, J. A., Smith, M. G., & Ward, M. J. (1989). *Astrophys. J.*, 342, 666.
[18] Schmidt, M. & Matthews, T. A. (1964). *Astrophys. J.*, 139, 781.
[19] Nishihara, E., Yamashita, T., Yoshida, M., Watanabe, E., Okumura, S. I., Mori, A., & Iye, M. (1997). *Astrophys. J.*, 488, L27.
[20] Kriss, G. A., Davidsen, A. F., Zheng, W., Kruk, J. W., & Espey, B. R. (1995). *Astrophys. J.*, 454, L7.

Here, the model developed in Section II presents a systematical explanation for the *redshift* variation of lines.

B. Redshift versus the number of massive particles

The *Doppler effect* cannot explain the *limb effect*, the signal *redshift* of Pioneer 6, and the large *redshift* of *quasars*. These "irregular" *redshifts* will be explained with the new model in the following.

1. *Redshifts on the solar disc*

The Sun is a special star for human beings. It is the only star we can observe in detail since it is near the Earth. Although the *limb effect* was discovered more than a century ago[3], it could not be explained perfectly before by any theories. In the following, it will be explained by the new model described in Section II.

Assis[5] discussed the *limb effect* and concluded [in 1993] that the *"tired-light" theory* provided a satisfactory explanation.

[5] Assis, A. K. T. (1993). *Progress in New Cosmologies: Beyond the Big Bang*, edited by H. C. Arp, C. R. Keys, and K. Rudncki, Plenum Press, New York, p. 153.

He suggested that the *redshift* was due to the interaction of light with the atmosphere of the Sun during its passage through it.

Adam[21] presented a set of varying *redshifts* of 14 lines on the *solar disc* at seven positions from the center to the *limb*, with an obvious *limb effect*.

[21] Adam, M. G. (1948). *Mon. Not. R. Astron. Soc.*, 108, 446.

LoPresto *et al.*[3] observed infrared oxygen triplet absorption lines on the *solar disc* at seven positions from the center to the limb, and obtained a result showing a *limb effect* weaker than that found by Adam.

[3] LoPresto, J. C., Schrader, C., & Pierce, A. K. (1991). *Astrophys. J.*, 376, 757.

The deviation of the two data sets from each other indicates that there must be some factor on the solar surface affecting the main trend of the *limb effect*. This factor is most probably the massive movement of *material upward* (MU) in the convection zone and in the *photosphere* of the Sun. The MU decreases the *redshift* by different degrees for *emission* or *absorption* lines produced at different depths. The two sets of observed lines were emitted from different depths in the Sun's atmosphere so that different changes were caused on the main trend of the *limb effect*.

For a given λ_0 included in the coefficient k, Eq. (7) becomes

$$Z = \exp(kN + u) - 1. \tag{11}$$

The *redshift* is then a function of N, the *number* of massive particles that the *photon* encounters in its journey from the Sun to the observer on the Earth.

The atmosphere of the Sun is a thick layer of hot gaseous massive particles covering the Sun. Along the line of sight, the thickness of the layer increases from the center to the *limb* of the *solar disc*. Figure 6 shows the thickness M of the gas layer along the line of sight, on supposing that all lines of sight are parallel. And, I disregard the variation of the density of the gas layer which is assumed to be of thickness L.

From the relations …

…

The *redshift* curve coincides with all the data points as shown in Fig. 7.

…

2. Signal redshift of Pioneer 6

When Pioneer 6 on its orbit at the other side of the Sun to Earth was approaching the Sun in November 1968, the signal from it to the Earth gave an additional *frequency shift*, or *redshifted*, by days. Chastel and Heyvaerts[22] showed the *frequency shifts*; Merat *et al.*[23] showed the corresponding *redshifts*.

[22] Chastel, A. A. & Heyvaerts, J. F. (1974). *Nature*, 249, 21.

[23] Merat, P., Pecker, J. C., & Vigier, J. P. (1974). *Astron. Astrophys.*, 30, 167.

Merat *et al.*[23] analyzed the reasons for the phenomena, and concluded that the data "strongly favor the existence of a new *redshift* cause at work in the Sun's vicinity."

The signal *redshift* of Pioneer 6 can be explained by the new model, like the explanation to the *limb effect*. I simplify the condition of the atmosphere of the Sun to a certain depth with a constant density for convenience of analysis. After the signal of Pioneer 6 began to penetrate the atmosphere of the Sun, the path of the signal was nearer to the *limb* of the *solar disc* from day to day; the signal would gradually go through a longer traveling distance (TD) in the atmosphere of the Sun. So, the signal of Pioneer 6 would meet and interact with more massive particles of the atmosphere of the Sun, and got *redshifted* more heavily by days.

…

3. A possible explanation for the large redshift of quasars

The *quasars* are queer for their large *redshifts* as their nature is not clear.

> [A *quasar* is an extremely *luminous active galactic nucleus*
> (AGN). The emission from an AGN is powered by a
> *supermassive black hole* with a mass ranging from millions to
> tens of billions of solar masses, surrounded by a gaseous
> accretion disc. Gas in the disc falling towards the *black hole* heats
> up and releases *energy* in the form of *electromagnetic radiation*.
> The radiant energy of quasars is enormous; the most powerful
> quasars have *luminosities* thousands of times greater than that of
> a galaxy such as the Milky Way. The *redshifts* of quasars are of
> cosmological origin.]

They might be a kind of celestial object with a much thicker and denser
layer of atmosphere, i.e., gaseous massive particles, than normal stars.
*On this assumption, the new model can provide a possible explanation
for the large redshifts of the quasars.*

The main part of the *cosmological redshift* (CR) is the *"tired-light"-
redshift* induced by the interaction of *photons* with massive particles. The
greater the *number* of massive particles that a *photon* encounters, the
larger the *redshift* that is induced. Light emitted from a *quasar* has to
penetrate through the atmosphere of the *quasar* so that the light from the
quasar should be *redshifted* much more heavily than that from a normal
star at the same distance to the Earth. Thus, the *quasar* does not need to
be located so far away as indicated by the large *redshift* with
interpretation of *Doppler effect*.

IV. DISCUSSION

Hubble[24] discussed the two different interpretations of *redshift*, the
Doppler shift, and the *energy-loss shift*; then, he said: "Thus the familiar
interpretation of *redshifts* as *velocity-shifts* leads to strange and dubious

conclusions; while the unknown, alternative interpretation leads to conclusions that seem plausible and even familiar."

[24] Hubble, E. (1937). *The Observational Approach to Cosmology.* Oxford University Press, Oxford.

Reber[2] described the history of the *Doppler effect*. The *Doppler effect* was proposed and first confirmed experimentally in the field of acoustics, and then confirmed observationally in the field of optics using *Fraunhofer lines* from the Sun in the nineteenth century. Then, the *Doppler Effect* was effectively applied to study the motion of double stars, the motion of the Sun in the Milky Way, and the rotation of our galaxy. In these situations, there are *redshifts* and *blueshifts* existing together, and *the magnitude of frequency shift has no relation to the distance*. Afterwards, the *Doppler effect* was applied to interpret the *cosmological redshift* (CR) when human observational ability extended into the intergalactic space. Although the *Doppler effect* had been successfully applied within our galaxy, the situation of CR is different since there are only *redshifts* but no *blueshifts*, and *the magnitude of the cosmological redshift (CR) is related to the distance the light traveled*, which is not consistent with the definition of the *Doppler effect*. That makes the interpretation of *Doppler effect* to the *cosmological redshift* (CR) being suspect, just as Reber[2] said, "Clearly the interpretation of these *spectral shifts* as representing relative motion was dubious."

At the time of Hubble's discovery, the *Doppler effect* was the only clear and simple concept available to interpret the discovery, for which the "warriors" of that era had no other choice. That situation had come from an assumption, as pointed out by Reber:[2] "The assumption is that *intergalactic space* is a void". We know, today, that there is *intergalactic medium* (IGM) in the intergalactic space. If we assume that the IGM is approximately distributed evenly in the intergalactic space, the *number* of massive particles of IGM on the traveling-way of light should be proportional to the distance, from which comes the basic thought of *"tired-light" hypothesis* of Zwicky.[4]

[In 1993,] Assis[5] and [1995,] Assis and Neves[25] discussed a model of the universe, which was developed by scientists, such as Regener, Nernst, Finlay-Freundlich, Max Born, and Louis de Broglie.

[25] Assis, A. K. T. & Neves, M. C. D. (1995). *Astrophys. Space Sci.*, 227, 13.

The model describes a steady universe infinite in space and time. In this universe the cosmological redshift (CR) is interpreted by use of the "tired-light" theory, and the new mechanistic model developed here will provide a basis for the *"tired-light" theory.*

Hubble[24] had said, "Light may lose *energy* during its journey through space, but if so, we do not yet know how the loss can be explained." Here, I am trying to answer the question of "how the loss can be explained." But, at present stage, the *"tired-light" theory* cannot explain the process of *energy loss* in detail. It needs to be developed by further studies. Preliminary tests have shown that the new model agrees with the observational data in the literature cited. More data checks and observational tests are needed for further evidence in support of the new model.

ACKNOWLEDGMENTS

I am indebted to Dr. Zhi-Qiang Shen, Dr. Bi-Ping Gong, Dr. Na Wang, Dr. Xiang Liu, Dr. Jian-Jun Zhou, and Dr. ZhiFu Gao for helpful discussions. I would like to thank Professor A. K. T. Assis and Mr. S. Crothers for their valuable advice, which they have given to me in their inspiring communications with me. This work is the result of a long period of reasoning and of wandering thoughts over the last 10 years.

I am glad to express my thanks to XAO-CAS for providing me with an opportunity to work in the field of astronomy, which relates directly to this work. ...

Lerner, E. J., Falomo, R., & Scarpa, R. (2014). UV surface brightness of galaxies from the local universe to z ~ 5.

International Journal of Modern Physics D, 23, 06, 1450058; https://doi.org/10.1142/S0218271814500588; also at https://arxiv.org/pdf/1405.0275.

This is an important article that demonstrates errors in previous claims that have been made that *the Tolman test provides compelling evidence against a static model for the universe.* This was reconsidered by adopting a *static Euclidean universe* (SEU) with *a linear Hubble relation at all z* (which is not the *standard Einstein–de Sitter model*), resulting in a relation between *flux* and *luminosity* that is virtually indistinguishable from the one used for *ΛCDM models*. Based on the analysis of the UV *surface brightness* (SB) of *luminous* disk galaxies from HUDF and GALEX datasets, reaching from the local universe to z ~ 5, it was shown that the *surface brightness remains constant as expected in a static universe.*

In particular, it shows that the conclusions in Lubin & Sandage *are not supported by the data* for two main reasons: (1) for the *static scenario,* Lubin and Sandage set the distance to d = (c/H_0) ln(1 + z), *which is valid only for the Einstein-de Sitter static case.* This is not the cosmology being tested where the *Hubble relation* is hypothesized to be d = cz/H_0 at all *redshift.* The conversion factors to transform arc seconds to pc in the *non-expanding model* are therefore different; (2) *the local sample includes only first rank cluster galaxies, while the high-z sample includes about 20 normal galaxies in each of three different clusters.* This means that their distant galaxies are on average smaller and less *luminous,* and therefore are not directly comparable to local ones because of the well-known *absolute magnitude-SB relation. It concludes that far from disproving a non-expanding cosmology, data by Lubin and Sandage agree very well with predictions for a static Euclidean universe.*

Abstract

The *Tolman test for surface brightness* (SB) *dimmin*g was originally proposed as a test for the *expansion of the universe*. The test, which is independent of the details of the assumed cosmology, is based on comparisons of the SB of identical objects at different cosmological distances. *Claims have been made that the Tolman test provides compelling evidence against a static model for the universe.* In this paper we reconsider this subject by adopting a *static Euclidean universe* (SEU) with a linear *Hubble relation* at all z (which is not the *standard Einstein–de Sitter model*), resulting in a relation between *flux* and *luminosity* that is virtually indistinguishable from the one used for *ΛCDM models*. Based on the analysis of the UV SB of *luminous* disk galaxies from HUDF and GALEX datasets, reaching from the local universe to z ~ 5, we show that the SB *remains constant as expected in a static universe.*

A re-analysis of previously published data used for the *Tolman test* at lower *redshift*, when treated within the same framework, confirms the results of the present analysis by extending our claim to elliptical galaxies. We conclude that available observations of galactic *surface brightness* (SB) are consistent with a *static Euclidean universe* (SEU) model.

We do not claim that the consistency of the adopted model with SB data is sufficient by itself to confirm what would be a radical transformation in our understanding of the cosmos. However, we believe this result is more than sufficient reason to examine this combination of hypotheses further.

---

1. Introduction

As Tolman[1,2] demonstrated, the dependence of the bolometric *surface brightness* (SB) of identical objects as a function of *redshift* z is independent of the specific parameter of the adopted cosmology, e.g.,

189

Hubble constant, dark matter Ω_M and *dark energy* Ω_Λ content of the Universe.

[1] Tolman R. C. (1930). *Proc, Natl. Acad. Sci.*, 16, 511.
[2] Tolman R. C. (1934). *Relativity Thermodynamic and Cosmology*. Oxford University Press, Oxford.

For this reason, the comparison of the *surface brightness* of similar objects at different distance was seen as a powerful tool to test for the *expansion of the Universe*. In fact, in any *expanding cosmology*, the *surface brightness* (SB) is expected to decrease very rapidly, being proportional to $(1+z)^{-4}$, where z is the *redshift* and where *surface brightness* (SB) is measured in the bolometric units (VEGA-magnitudes/arcsec^{-2} or erg sec^{-1}cm^{-2}arcsec^{-2}). One factor of $(1+z)$ is due to *time-dilation* (*decrease in photons per unit time*), one factor is from the *decrease in energy carried by photons*, and the other two factors are due to *the object being closer to us by a factor of (1+z) at the time the light was emitted* and thus having a *larger apparent angular size*. (If *AB magnitudes* or *flux densities* are used, the dimming is by a factor of $(1 + z)^3$, while for space telescope magnitudes or flux per wavelength units, the dimming is by a factor of $(z + 1)^5$).

[The *AB magnitude system* is an *astronomical magnitude* system. Unlike many other magnitude systems, it is based on *flux measurements* that are calibrated in absolute units, namely *spectral flux densities*.]

By contrast, in a *static (non-expanding) Universe*, where the *redshift* is due to some physical process other than expansion (e.g., *light-aging*), the SB is expected to dim only by a factor $(1 + z)$, or be strictly constant when *AB magnitudes* are used.

In the last few decades, the use of modern ground-based and space-based facilities have provided a huge amount of high-quality data for the high-z Universe. The picture emerging from these data indicates that galaxies evolve over cosmic time. The combination of cosmological effects with the evolution of structural properties of galaxies makes the *Tolman test* more complicated to implement because of the difficulty in disentangling two types of effects (*cosmology* and *intrinsic evolution*). In spite of this complexity, various authors have attempted to perform the *Tolman test*[3,4,5], most reaching the conclusion that the *Tolman test* ruled out the *static Universe model* with high confidence.

[3] Pahre, M. A., Djorgovski, S.G., & de Carvalho R. R. (1996). *Astrophys. J.*, 456, L79.

[4] Lori M. Lubin, L. M. & Sandage, A. (2001). The Tolman Surface Brightness Test for the Reality of the Expansion. IV. A Measurement of the Tolman Signal and the Luminosity Evolution of Early-Type Galaxies," *Astron. J.*, 122, 1084-103.

[5] Sandage A. (2010). *Astron. J.*, 139, 728.

In this paper we present a new implementation of the *Tolman test* based on a comparison of the UV *surface brightness* of a large sample of disk galaxies from the local Universe to z ~ 5 as well as a critical re-analysis of previously-published data. Preliminary reports of this work were presented by Lerner[6,7].

[6] Lerner, E. J. (2006). *AIP Conference Proceedings*, 822, 60.

[7] Lerner, E. J. (2009). *ASP Conference Series*, 413, 12.

Consistent with those preliminary reports and *contrary to earlier conclusions by other authors*, we here show that *the surface brightness of these galaxies remains constant over the entire redshift range explored*. Based on these observations, *it is therefore not true that a static Euclidean Universe can be ruled out by the Tolman test*.

2. The adopted cosmology.

Since the *surface brightness* (SB) of galaxies is strongly correlated with the *intrinsic luminosity*, for a correct implementation of the *Tolman test* it is necessary to select samples of galaxies at different *redshifts* from populations that have on average the *same intrinsic luminosity*. To do this, one is forced to adopt a relation between z and distance d in order to convert *apparent magnitudes* to *absolute magnitudes*. In this paper we are testing a *static cosmology* where *space is assumed Euclidean* and the *redshift* is due to some physical process *other than expansion*. For this study, we adopt the simple hypothesis that the relationship $d = cz/H_0$, well-assessed in the *local Universe*, holds for all z. It should be noted that *this cosmological model is not the Einstein-De Sitter static Universe* often used in literature.

The choice of a *linear relation* is motivated by the fact that the *flux-luminosity relation* derived from this assumption is remarkably similar numerically to the one found in the *concordance cosmology*, the distance modulus being virtually the same in both cosmologies for all relevant *redshifts*.

> [The term '*concordance model*' is used in cosmology to indicate the currently accepted and most commonly used cosmological model. The *Lambda-CDM, Lambda cold dark matter*, or *ΛCDM model* is a mathematical model of the *Big Bang theory* with three major components:
> 1. a *cosmological constant*, denoted by lambda (Λ), associated with *dark energy*;
> 2. the postulated *cold dark matter*, denoted by CDM;
> 3. *ordinary matter*.
>
> It is referred to as the *standard model* of *Big Bang cosmology* because it is the simplest model that provides a reasonably good account of:
> - the existence and structure of the *cosmic microwave background*;
> - the *large-scale structure* in the distribution of galaxies;
> - the *observed abundances* of hydrogen (including deuterium), helium, and lithium;
> - the *accelerating expansion of the universe* observed in the light from distant galaxies and *supernovae*.

The model assumes that general relativity is the correct theory of gravity on cosmological scales. It is consequently inconsistent with New Physics. It emerged in the late 1990s as a *concordance cosmology*, after a period of time when disparate observed properties of the universe appeared mutually inconsistent, and there was no consensus on the makeup of the *energy density* of the universe.]

This is shown in Fig. 1 where the two relations are compared to each other and, in Fig. 2, to *supernovae* type Ia data. Up to *redshift* 7, the apparent magnitude predicted by the *simple linear Hubble relation* in a *Static Euclidean Universe* (SEU) is within 0.3 magnitude of the *concordance cosmology* prediction with $\Omega_M = 0.26$ and $\Omega_\Lambda = 0.74$. The fit to the actual *supernovae* data is statistically indistinguishable between the two formulae.

In this particular framework the *bolometric luminosity* L and the *flux* F from a source are related by the relation $F = L/[4\pi d2(1+z)]$, where the factor $(1+z)$ takes into account *energy losses* due to the *redshift*.

[*Bolometric luminosity* is an astronomical body's total power output across all *electromagnetic radiation wavelengths*.]

When using *flux per unit frequency*, that is *AB magnitudes*, this relation further simplifies to $F = L/(4\pi d2)$. Therefore, the *absolute magnitude* M can be derived from the *apparent magnitude* m (in the *AB system*) using the relation: $M - m = 5 - 5 \, \text{Log} \, (cz/H_0)$.

Under the assumption of a *static Universe* the *true size* R and the *apparent size* r of an object are linked by the standard relation $r = R/d$, where d is the *distance* and r is in radians. The average *surface brightness* μ (in *magnitude*) of a galaxy becomes $\mu = m + 2.5 \, \text{Log} \, (2\pi r^2)$, where m is the *total apparent magnitude*, r the *radius*. As the *radius* does not depend on z, from this definition it follows that the *apparent surface brightness* is expected to get dimmer as m, that is $\mu \sim (1 + z)^{-1}$ when using standard *VEGA magnitudes*, or remain constant when using *AB magnitudes*.

> [The *Vega magnitude system* uses the star *Vega* (α Lyrae) as the reference star. In this system, *Vega* has *magnitude* 0 at all *frequencies*. The logarithmic *magnitude* scale mimics the intensity sensitivity of the human eye. Other stars are measured relative to *Vega's magnitude*. The modern *magnitude* system measures the *brightness* of stars, not their *apparent size*.]

In the following we use *AB magnitudes*. In applying this *linear relation* between z and d, we are not here proposing any physical model that would produce such a relation — we simply extrapolate the local properties of the Universe to see whether they are consistent with the *surface brightness* data.

3. The Samples definition.

At present, the best data set for studying the properties of objects in the distant Universe is the *Hubble Ultra Deep Field* (HUDF) 10, which is a set of 4 images obtained with the *advanced camera for survey* (ACS) in the B, V, I, and z bands down to an unprecedentedly faint apparent magnitude ($m_{AB} \sim 29$). To avoid large and uncertain k-corrections, the *surface brightness* (SB) must be compared as much as possible at the same *rest frame wavelengths* for all objects. To satisfy this condition and properly compare galaxies up to z~5, we have chosen two reference ultraviolet bands, namely the FUV (1550 Å) and NUV (2300 Å) bands as defined by the GALEX satellite, enabling the creation of 8 pairs of samples matched to the HUDF data.

To minimize the effects of k-correction, the *redshift* range covered by each GALEX-HUDF pair was set requiring a maximum difference of 10% between the central *rest wavelength* determined by the GALEX and ACS filters. Moreover, to avoid biasing the comparison of data obtained with telescopes having different resolutions, we also require that the *minimum measurable physical size* of galaxies rm is the same, in each pair of samples, for GALEX (low z) and HUDF (high z). We have determined the *minimum measurable angular radius* of galaxies, θ_m, for each of the telescopes by plotting the *abundance of galaxies* (with stellarity index < 0.4) vs. *angular radius* for all GALEX MIS3-SDSSDR5 galaxies and for all HUDF galaxies and determining the lower-cutoff *angular radius* for each. We took this cutoff to be the point at which the *abundance per unit angular radius* falls to 1/5 of the modal value. For GALEX this cutoff is at a *radius* of 2.4 ± 0.1 arcsec for galaxies observed in the FUV and 2.6 ± 0.2 arcsec for galaxies observed in the NUV, while for Hubble this cutoff is at a radius of 0.066 ± 0.002 arcsec, where the errors are the 1 σ statistical uncertainty. We averaged the NUV and FUV cutoffs to find the ratio of θ_mGALEX/θ_mHUDF to be 38 ± 3. In accord with our test model, with *minimum measurable physical radius* $r_m \sim z\theta_m$, we chose pairs of samples so that the ratio of mean z in the HUDF sample to mean z in the GALEX sample is also as close as possible to ~38. Thus r_m, assuming the model, is the same for each member of the pair of samples.

In order to avoid effects due to the *luminosity* of galaxies, we limited objects in the samples to a narrow range of *absolute magnitude* M: $-17.5 < M < -19.0$, matching the mean *absolute magnitude* of each pair down to 0.02 mag, in such a way as to maximize the total number of galaxies in the pair. These are the brightest galaxies that are present in both GALEX and HUDF samples. Because galaxy size increases somewhat with *absolute luminosity*, these are also the galaxies most easily resolved and measured by both instruments. These UV data have the important advantage of being sensitive only to emissions from very young stars. Therefore, we are in no sense looking at progenitors of GALEX galaxies, but rather at galaxies whose stellar populations are comparable in age. By analogy we are looking at populations of "babies" at different epochs in history, not comparing younger and older adults born at the same time. The important question of the comparability of the GALEX and HUDF samples is dealt with in greater detail in Section 5.3.

195

Finally, we restricted the samples to disk galaxies with *Sersic number* < 2.5 so that *radii* could be measured accurately by measuring the slope of the exponential decline of *surface brightness* (SB) within each galaxy. This measurement technique, using the slope of SB to determine *radius*, eliminates errors that can be introduced by measuring the *radius* at some arbitrarily determined *isophote*. For the GALEX sample, we measured *radial brightness* profiles and fitted them with a *Sersic law*, finding that nearly all these bright UV galaxies, as expected, had *Sersic number* < 2.5. For HUDF, we used the *Sersic number* provided in the HUDF catalog 11. We also used the HUDF and GALEX catalogs to exclude all non-galaxies. The properties of the selected galaxies are summarized in Tables 1, 2 and 3.

4. Determination of redshift, radius and magnitude of galaxies.

For the HUDF dataset, the *redshift* was based on the HUDF photometric catalogs. These catalogs contain photometric measurements for each galaxy in the B, V, I, z, H and J bands. Each galaxy has a photometric *redshift*, estimated by two methods: *Bayesian Probability* (BPZ) and *Maximum Likelihood* (BML). Coe *et al*[11] report that a comparison of BPZ with spectroscopic *redshifts* in the small sample where they are available indicates that, except for a few outliers, BPZ *redshifts* are accurate to 0.04.

[11] Coe D., Bentez N., Snchez S. F., Jee M., Bouwens R., & Ford H. (2006). *Astronomical Journal*, 132, 926.

To eliminate outliers, we have chosen to use the difference between BML and BPZ *redshifts* as an indicator of the reliability of BPZ *redshifts*, retaining only sources for which the two *redshifts* differ by less than 0.5. For GALEX, we limited our samples to galaxies with *spectroscopic redshift* derived from cross-correlating the MIS3 with data from the SLOAN Digital Sky Survey (SDSS) Data Release 5.

To measure *total flux* and *half-light radius*, we extracted the average *surface brightness* profile for each galaxy from the HUDF or GALEX images. The *apparent magnitude* of each galaxy is determined by measuring the *total flux* within a fixed circular aperture large enough to accommodate the largest galaxies, but small enough to avoid contamination from other sources. To choose the best aperture over which to extract the radial profile, for each sample we compared average *magnitudes* and average *radii* as derived for a set of increasingly large apertures. We then defined the best aperture as the smallest for which average values converged. We found that these measurements are practically insensitive to the chosen aperture above this minimum value.

Finally, to determine *scale-length radius*, we fitted the *radial brightness* profile with a disk law excluding the central 0.1 arcsec for *Hubble Space Telescope* (HST) and 5 arcsec for GALEX, which could be affected by the *PSF smearing*. Given the *magnitude* and *radius*, the *surface brightness* (SB) is obtained via the formulae in Section 2. A direct comparison between our measurements and those in the i band HUDF catalogue[11] show no significant overall differences. The SB for all selected galaxies is shown in Figure 3 plotted against *redshift*.

5. The Tolman test

5.1 *Comparison of surface brightness*

To perform the *Tolman test*, for each pair of data sets we compute the difference of average *surface brightness* between the low and high z dataset. These results are shown in Figure 4 and Table 5. The difference of *surface brightness* (SB) between the pairs is always very small and no obvious trend depending on the *redshift* is apparent. The mean SB difference of all samples taken together, weighted by the number of galaxies in each pair, is 0.027 ± 0.033 mag/arcsec^2 (1σ statistical uncertainty). A linear fit of SB differences with the $<z>$ of the HUDF samples yields a slope of ΔSB on $<z>$ of 0.04 ± 0.06 mag/arcsec^2 (coefficient of correlation 0.28) and therefore is consistent with no correlation. Therefore, these data are fully consistent with the *surface brightness* (SB) being constant in the *redshift* range explored.

197

We investigated whether the different resolutions of the two telescopes could bias the comparisons because different portions of the population distribution of SB are excluded as unresolved galaxies in the two (low-z and high-z) samples of each pair. If, for example, in the HUDF samples most galaxies are resolved, while in the GALEX sample most galaxies are unresolved, the underlying populations of objects may have very different *average surface brightness*, <SB>, even if the <SB> of the resolved samples are the same. In this respect we point out that in the adopted *Euclidean model*, the GALEX and HUDF samples probe the same range of galaxy *radius distribution*.

To quantify and eliminate these possible biases we performed the *Tolman test* including all galaxies, both resolved and unresolved. To do this we made two justifiable assumptions. First, we assumed that the proportion of disk galaxies that were unresolved was the same as the proportion for all galaxies that were unresolved. That enabled us to estimate the number of unresolved galaxies for each sample. We computed the ratio of the number of unresolved galaxies (those with *stellarity* > 0.4) to the number of resolved galaxies for all galaxies within the *redshift* and *absolute magnitude* limits defined by each of the sub samples that we have selected (see Table 4). This comparison shows that there is no significant difference between the proportion of unresolved galaxies in the GALEX and HUDF datasets. Note also that, except for the HUDF FUVz sample, (where two-thirds of the galaxies are resolved), the resolved galaxies greatly outnumber the unresolved ones and 75% of all galaxies in each of the bins are resolved. This gives a preliminary indication that our analysis is not significantly affected by any biased population of galaxies due to the different resolution of the telescopes. It is also worth noting that this check is *almost independent of cosmological assumptions* because *redshift* is an observed quantity and *absolute magnitude* is close in the two models considered.

Second, we assumed that the *surface brightness* (SBs) of all the unresolved galaxies were brighter than that of the median galaxy of the population. We then determined the median SB galaxy within each sub-sample, by ranking all measured *surface brightness* (SBs) in the sample and including the estimated number of unresolved galaxies as being below (in value) the median. We then compared the median SB of the GALEX and HUDF samples within each pair as we did with the mean of the resolved galaxies. For a Gaussian distribution, or any symmetrical distribution, the mean and median values (of the whole population, resolved and unresolved) should be equal within statistical errors. The results are shown in Figure 5 and are compared with the mean SB results in Table 5. The mean of all eight differences, weighted by the number of galaxies in the pairs, is -0.017 ± 0.05 mag/arcsec^2, the slope of ΔSB on z(HUDF) is -0.08 ± 0.05 mag/arcsec^2 with a correlation of 0.53, insignificant for 8 points even at a 5% level. This is all still completely consistent with zero difference in *surface brightness* (SB) between high-z and low-z samples and with no dependence of SB on z, in accord with *Tolman test* predictions for a *static Euclidean Universe*.

We can use the median *surface brightness* (SB) value to obtain an estimate of the variance within each sample by measuring the variance of all galaxies with SB more (in value) than the median SB. These variances are used to calculate the error bars (expected variance of sample median or sample mean) in Figures 4 and 5. With these variances, we can determine if the variation of the ΔSB, measured either way, is greater than that expected purely from random variation in the samples. Using a chi-squared test, we see that for both methods, chi-squared is well below the 5%-probability limit of 14.1 for 7 degrees of freedom, being 9.0 using the *mean SB method* and 12.4 using the *median SB method*. Thus, the null hypothesis that the differences are due only to the variability of the samples is accepted. The variances expected in sample medians are in fact somewhat underestimated since we do not take into account errors created by uncertainty in the actual number of unresolved galaxies. We thus see that both versions of the Tolman test, either ignoring or taking into account the unresolved galaxies, *are both entirely consistent with a static Euclidean Universe prediction of no variation in SB and entirely consistent with each other.* Indeed, overall SB results for the GALEX and HUDF samples differ from each other by less than the statistical uncertainty of 0.03-0.05 mag/arcsec^2, a strikingly close agreement.

Finally, we have checked, by visual inspection of galaxies in the sample, that removing objects exhibiting signatures of interaction or merging do not change our conclusions. The selection of galaxies with disturbed morphology was performed by an external team of nine amateur astronomers evaluating the NUV images and *isophote* contours of all NUV-sample galaxies. Each volunteer examined the galaxies and only those considered unperturbed by more than 5 people were included in a "gold" sample. Although this procedure reduces the size of the sample, there is no significant difference of the SB-z trend.

5.2 *Is there a bias for size or surface brightness?*

We examined whether our results could be the result of an implicit selection for either *surface brightness* or, equivalently, for the *radius* of galaxies of a given *intrinsic luminosity*. The limited *angular resolution* of the observations imply that there is a *minimum angular radius* for resolved galaxies and thus a *maximum surface brightness* (SB) for galaxies of a given M and z. As well, there is a limit on the dimmest SB that each telescope can observe, which puts a minimum on the SB that can be included in the sample. Together, these limits inevitably restrict the measured SB of any galaxy sample within a window of minimum and maximum SB. Are our results biased by these limits, simply reflecting the range of this window, and are we implicitly thus selecting for a narrow range of SB or angular size?

We can answer unequivocally that this "windowing" does not affect our results and that we have not imposed an implicit selection on *radius* or *surface brightness* (SB). First, we are including for the evaluation of the *median SB* ALL observed galaxies in the defined M and z ranges, whether or not they are resolved. Thus, as more than half of all galaxies in the sample and in each sub-sample are resolved, the value of the *median SB* is not affected by the maximum observable SB imposed by the telescope resolution.

Second, we note that, for the *very luminous* galaxies that we have chosen, the low-SB limits of both the HUDF and the GALEX MIS surveys are sufficiently far from the distribution of SB actually observed, in other words, the "window" is sufficiently wide, that these limits also have no effect on our median SBs. In figure 6 we show that for both GALEX and HUDF, the SB distribution for galaxies with $-16 < M < -17$ does not even begin to decrease until 28 mag/arcsec^2, dimmer than all galaxies in both of our samples with $-17.5 < M < -19$ and more than 2 sigma away from the peak of the distribution for our samples. Thus, because the bright galaxies we selected are large enough to be well resolved and bright enough not to be missed even at their largest, the measurement of the median SB is not affected by the "window" effect described above.

5.3 Sensitivity of the results

In implementing the *Tolman test*, we have taken care to match (using the SEU model) the *linear resolutions*, the *rest wavelengths* and the *absolute magnitudes* of the samples. How sensitive is the test to the accuracy of these matches? A comparison of FUV with NUV SB at the same z and M using the GALEX samples shows that *surface brightness* (SB) in the *wavelength* range covered appears insensitive to λ, with a slope of only 0.35 mag/arcsec^2 of SB on log λ .Thus, the 8% variance we allowed in λ only results in SB difference of 0.01 mag, much less than statistical errors. Similarly, the slope of SB on the log of the resolution in the GALEX sample is 2.2 ± 0.2 mag/arcsec^2, so an error of 5% in the ratio of resolutions of samples (or in determining the ratio of resolutions between HUDF and GALEX) will produce a change in SB of 0.05 mags, the same as the statistical error. Thus, ratios in *angular resolution* in the range of 36-40 would not have a statistically significant effect. By comparison, a choice of *cutoff* in determining effective resolution, anywhere from ½ to 1/10 the modal value, would vary the ratios by less than ± 2%. Finally, the slope of SB on M is close to 1.0, so we did need to keep the <M> of the samples close to each other, as it would only take a change of 0.07 mag in to produce the same change in <M>. As can be seen from Table 1, the maximum difference is only 0.02 mag in <M>. Thus, we conclude that our results are robust within the statistical errors.

5.4. *Effects of colors*

Since different stellar populations of low and high *redshift* galaxies might produce some systematic effects in the derived *surface brightness*, we have investigated the NUV-g colors of the selected galaxies, colors that are sensitive to the age of the stellar population.

We note that the NUV-g colors of the GALEX and HUDF samples are significantly different from each other, even if similar when compared with all galaxies, with the HUDF sample 1.3 mag bluer. However, both samples have colors typical of stellar populations with ages <1 Gyr, far separated from old, inactive galaxies. For our purposes here the key point is that for these *very luminous* galaxies there is no correlation at all between SB and NUV-g color, so the differences in color between HUDF and GALEX samples have no effect on our results.

To test for such correlations with the GALEX samples, we must avoid selection biases introduced by the SDSS *redshift selection algorithm*. The SDSS selection[12] eliminates galaxies with r-band SB > 23 mag/arcsec^2.

[12] Strauss, M. A. *et al.* (202). *Astronomical Journal*, 124, 1810.

This means that it also eliminates galaxies with blue NUV-g colors and relatively dim NUV SB. SDSS also eliminated galaxies with r-band radius <2 arcsecs, setting an effective SB minimum to r-band of r + 3.5 mag/arsec^2. This similarly eliminates galaxies with red NUV-g colors and relatively bright NUVSB. In both cases, the selection limits tend to create a spurious correlation of NUV-g and NUV-SB.

We can minimize such biases and obtain a true correlation of NUV-g and SB by limiting our test sample to the closest galaxies, with z < 0.05. These galaxies are close enough, within 200 Mpc, that none are affected by the minimum r-band radius cutoff. In addition, for near-by galaxies, the SDSS SB selection limit is somewhat relaxed. So, we limit the GALEX samples to the NUVB and NUVV samples. We find that for the GALEX samples there is no correlation between *surface brightness* (SB) and NUV-g color even at the 5% level. For the entire HUDF sample, we find the same lack of correlation. *Thus, differences in color between GALEX and HUDF samples have no effect on the SB comparison.* This is not particularly surprising. Since we have limited the samples in *luminosity*, the lack of change in SB simply means that there is no significant change in *radius* for these large, actively star forming galaxies with respect to the age of the stellar populations. We noted above that, for these bright galaxies, the slope of SB on M is close to 1, which means that *radius* does not vary greatly with *absolute luminosity* either, so it is not surprising that it does not vary much with color. Finally, we confirm that selection biases noted in the SDSS catalog affect only color-SB plots, not the overall <SB> of the samples, because there is no statistical difference between the comparison involving the NUVB and NUVV samples, which do not suffer from the selection, and the rest of the samples, which do.

5.5 *General remarks on analyses using size evolution*

In this paper we are examining the consistency of data on the *surface brightness* (SB) of galaxies using the *static Euclidean* (SEU) *model* with *redshift proportional to distance*. We therefore do not expect any evolutionary effects either in *size* or *luminosity*, in contrast to expectations in ΛCDM models. Not only are these in all cases galaxies whose UV radiation is dominated by young stellar populations, but in the *static Euclidean model* that we are testing the *mean density* of the universe remains a constant, so we expect no change with z among such young galaxies in *size* or in *virial radius* for a given *luminosity*.

The prediction of the *static Euclidean* (SEU) *model* that the *surface brightness* (SB) for a given *absolute luminosity* is constant with z is mathematically identical to the prediction that the *mean physical radius*

R of a population of galaxies with a given *absolute luminosity* is also constant with z. From the assumed linear relation of *redshift* with *distance*, this model also predicts that mean *angular radius* for such galaxies is *inversely proportional to z*. So, our SEU model demonstration that SB values are constant simultaneously demonstrates that mean R is also constant with z and that *angular radius* is inversely proportional to z, a conclusion also reached by Lopez-Correidora[13] for a lower-z sample.

[13] Lopez-Corredoira M. (2010). *Int. J. Mod. Phys.* D, 19, 245.

In this paper, we do not compare data to the *ΛCDM model*. We only remark that any effort to fit such data to ΛCDM requires hypothesizing a size evolution of galaxies with z. Mathematically, in order to fit the observed constancy of SB data, any expanding universe model must require that the *radii* of galaxies with constant *absolute luminosity* evolve exactly as $(1 + z)^{-1.5}$ in order to cancel out the $(1 + z)^3$ SB dimming. Conversely, theories that predict some other size evolution, such as $(1 + z)^{-1}$, will not fit the constant-SB data actually observed. Nor will this data be fit by any size evolution of the form $H(z)^{-a}$, where a is any constant and $H(z)$ is the *Hubble parameter* predicted by ΛCDM at a time corresponding to *redshift* z. In ΛCDM, $H(z) \sim (1 + z)^{1.5}$ at high z, but diverges greatly from this value at low z. For example, size evolution proportional to $H(z)^{-1}$, advocated by Hathi *et al*[14], among others, predicts a difference in SB between low-z and high-z samples of ~1 mag/arcsec^2, at z = 1, very far from the observations presented here, which show no difference in SB to within the statistical uncertainty of 0.05 mag/arcsec^2.

[14] Hathi N. P., Malhotra S. (2008). & Rhoads J. *Astrophs. J.*, 673, 686.

We leave to further work an examination of whether a size evolution that coincidentally cancels out SB dimming is physically plausible.

6. Previous implementations of the Tolman test revisited

We reconsider in this section previous works where it was concluded that a static Universe is ruled out by the Tolman test. We show that, when data are consistently analyzed within the framework of the *static cosmology* adopted here, they agree with the expectation of no dimming. We explicitly show the details of the reanalysis for two works. For other works considering the *Tolman test*[14,15] similar conclusions were found.

[15] Weedman D. W., Wolovitz J. B., Bershady M. A., & Schneider D. P. (1998). *Astron. J.*, 116, 1643.

6.1. Paper by Pahre, Djorgovski, and de Carvalho 1996

Pahre, Djorgovski, and de Carvalho[3] (PDdC hereafter) applied the *Tolman test* by studying the *surface brightness* (SB) of elliptical galaxies in 3 clusters up to z = 0.4. It was concluded that the data are in good agreement with the expectations for an *expanding Universe*, while the non-expanding model was ruled out at the better of 5 sigma significance level. *We demonstrate here that this is not the case.* To cope with the strong SB-radius correlation of elliptical galaxies, PDdC compared the SB at a fixed physical radius of 1 kpc computed for the *expanding Universe*, adopting H_0 = 75 km s^{-1} Mpc^{-1}, Ω_M = 0.2, Ω_Λ = 0. Unfortunately, they used the same *surface brightnesses* (SBs) computed for the *expanding case* to test also the *nonexpanding* one. Clearly, to make a fair test all the transformations from *apparent* to *physical sizes* must be properly computed for the *static model*, again *using the linear d-z relation.* When this is done, we see that the SBs used by PDdC refer to *physical radii* of 1.4 kpc at z = 0.23 and 1.7 kpc at z = 0.4 (see Table 6). Here for consistency with PDdC we use H_0 = 75 km s^{-1} Mpc^{-1}. Due to this effect at z = 0.4, an artificial SB dimming of ~0.5 magnitude is introduced. *This is fully responsible for the failure of the non-expanding model claimed by PDdC.* The corrected SBs are presented in Table 6. Note that *VEGA magnitudes* are used in this work, so a (1 + z) dimming is expected for the *static case*. This is shown in Figure 7 where the corrected data for the *static model* are compared with the predictions. We thus conclude that *when consistently analyzed these observations are in agreement with the expectation for a static Euclidean Universe.*

6.2. *Lubin and Sandage 2001*

In a series of four papers with final results in Lubin and Sandage[4] (LS01 hereafter), the *Tolman test* was applied comparing the *surface brightness* (SB) of *local early type galaxies* at *average redshift* < z > = 0.037, to the one of early type galaxies in three distant clusters, one at z = 0.75 and two at z = 0.9. Reinforcing an initial claim presented by Sandage and Perelmuter[16], it was concluded that the $(1 + z)^{-4}$ *surface brightness* dimming (LS01 used standard *VEGA magnitudes* and $H_0 = 50$ km s^{-1} Mpc^{-1} and $q_0 = 1/2$) was in agreement with observations *provided a significant amount of luminosity evolution was taken into account.*

[16] Sandage A., & Perelmuter J.-M. (1991). *Astrophys. J.*, 370, 455.

The nonexpanding scenario was ruled out at the 10σ confidence level. These very same data are also re-discussed by Sandage[5], reaching similar conclusions.

These conclusions are not supported by the data for two main reasons. The first one is that, for the *static scenario*, Lubin and Sandage set the distance to d = (c/H_0) ln(1 + z), *which is valid only for the Einstein-de Sitter static case.* This is not the cosmology we are testing here, where the *Hubble relation* is hypothesized to be d = cz/H_0 at all *redshift*. The conversion factors (presented in their Table 8) to transform arc seconds to pc in the *non-expanding model* are therefore different in our model. The second reason is that *the local sample includes only first rank cluster galaxies, while the high-z sample includes about 20 normal galaxies in each of three different clusters.* This means that their distant galaxies are on average smaller and less *luminous*, and therefore are not directly comparable to local ones because of the well-known *absolute magnitude-SB relation*.

LS01 presented *magnitudes* and *surface brightness* (SB) as derived for four different *Petrosian radii*, as defined by the η parameter, because a dependence on dimming with η was found. (By definition η is the difference in *magnitudes* between the *surface brightness* averaged over

206

a *radius*, to the *surface brightness* at that *radius*. Larger η corresponds to larger *radii*. For reference, the commonly used *half-light radius* corresponds to η = 1.4.) Contrary to LS01, in the *static cosmology* we are using, there is no difference in considering different values of η; thus, in the following we limit ourselves to η = 2, which was indicated by LS01 as the most appropriate for implementing the *Tolman test*. The distribution of *absolute magnitudes* derived in our *static cosmological model* for the local and distant samples is shown in Figure 8. As expected, there is a clear offset in *luminosity* between samples, *local galaxies being on average 1.5 magnitudes brighter than the distant ones*. Thus, to cope with the strong *SB-luminosity relation* we are forced to cut samples, considering only the region of overlap in *luminosity*, namely − 23.8 < M < − 22.7. We stress that the *luminosity* offset is the same on both the *expanding* and *static* scenario because, as pointed out in section 2, the *luminosity distance* is virtually the same in the two models. Thus, the limitation in *luminosity* range is legitimate and is not biasing the test. In Fig. 9 we plot the *surface brightness* (SB) of the 14 selected galaxies as a function of their size (computed in the *static scenario*). Using *Vega mag* a dimming of a factor (1 + z) is expected, thus data have been made brighter by this amount to be directly comparable. Within the intrinsic spread of the *SB-size relation*, the match between local and distance samples is good. There is only one clear outlier, the galaxy with the brightest *surface brightness*, which is the first entry in Table 2 of LS01. Excluding this outlier, we find a probability of 68% for the samples to be drawn from the same population. This was computed assuming a constant uncertainty of 0.15 mags on the SB, certainly an underestimation of the true uncertainty considering the large number of transformations required to convert observed quantities to the same rest frame system. Including the outlier galaxy decreases the probability but the samples remain statistically indistinguishable.

The samples overlap neatly in size, which is to say (in the non-expanding scenario) that galaxies of similar *luminosity* also have similar physical size and therefore, necessarily, the same SB. This is not the case in the expanding scenario, where samples do not overlap at all in *radius*, forcing Lubin and Sandage to extrapolate the local sample to small *radii* using data from Sandage and Perelmuter[16]. *We conclude that far from disproving a non-expanding cosmology, data by Lubin and Sandage agree very well with predictions for a static Euclidean universe.* This result effectively extends our own results, as discussed above, to early-type galaxies.

7. Conclusions

We find that the UV *surface brightness* of *luminous disk galaxies* are constant over a very wide *redshift* range (from z = 0.03 to z ~ 5). From this analysis *we conclude that the Tolman test for surface brightness dimming is consistent with a non-expanding, Euclidean Universe with distance proportional to redshift.* This result is also consistent with previously published datasets that were obtained to perform the *Tolman test* for a smaller *redshift* baseline when analysis of such data is done in a consistent system.

We stress that our analysis compared samples of galaxies that were matched for:
- mean absolute magnitude,
- rest-frame wavelength,
- minimum measurable physical radius,

thus removing the needs for complex and uncertain corrections. There is no implicit or explicit bias for *surface brightness* (SB) or *galaxy radius*.

We also emphasize that this matching of observations and predictions of the *non-expanding, Euclidean Universe* involves neither fitting of parameters nor any free variables. The simple prediction of constant *surface brightness* (SB), and equivalently, no *size evolution* in these young galaxies is consistent with all observations.

We have confirmed the constancy of *surface brightness* (SB) using two statistical methods for determining mean SB of a population, one of these methods including unresolved galaxies. A re-analysis of earlier data for elliptical galaxies, covering a different range of *redshift*, obtained with

208

different methods, and in different *wavelengths*, shows consistency with our results, thus extending the significance of the test.

The agreement of the *surface brightness* (SB) data with the hypotheses of a *non-expanding, Euclidean Universe* and of *redshift proportional to distance* is not sufficient by itself to confirm what would be a *radical transformation in our understanding of both the structure and evolution of the cosmos and of the propagation of light*. However, this consistency is more than sufficient reason to examine further this combination of hypotheses.

Mamas, D. L. (May, 2015). Cosmological redshift model now experimentally confirmed.

Physics Essays, 28, 2, 201-2; http://dx.doi.org/10.4006/0836-1398-28.2.201.

PhD Physics (UCLA). 4415 Clwr. Hr. Dr. N., Largo, Florida 33770.

Received: September 18, 2014.
Accepted: April 21, 2015.
Published online: May 6, 2015.

Abstract: A theoretical model [Mamas, D. L. (2010). *Phys. Essays*, 23, 326], which accounts for the *cosmological redshift* in a *static universe*, now has experimental confirmation. In this model, the *photon* is viewed as an *electromagnetic wave* whose *electric field* component causes oscillations in deep space *free electrons which then reradiate energy from the photon*, causing a *redshift*. Calculations from the model match well the anomalous *redshift* of Wernher von Braun's Pioneer 6 spacecraft.

I. INTRODUCTION

An explanation for the *cosmological redshift* with no expansion of space needs to be simple and must satisfy all required conditions such as avoiding the problem of blurring of images. Such a model has been presented[1], and this *wave particle model* now has found experimental confirmation.

[1] Mamas, D. L. (2010). *Phys. Essays*, 23, 326, see above.

In this *wave-particle* model, the *photon* is viewed as an *electromagnetic wave* which passes over *free electrons*, one at a time, in deep space. *Maxwell's equations* demand that a *free electron oscillates and reradiates energy*, necessarily at the expense of the *photon* which can then be expected to *red shift*. As previously reported in detail, *the wave-particle model accounts satisfactorily for the cosmological redshift*[1]. In accordance, Accardi *et al.* have demonstrated that it is permissible within the theory of *quantum electrodynamics* that a *photon, viewed as a wave in a Fermi Sea of free electrons*, lose *energy* to the *electron sea* and *redshift*[2].

[2] Accardi, L., Laio, A., Lu, Y. G. & Rizzi, G. (1995). *Phys. Lett.*, A 209, 277.

Finally, it should be noted that the simple *wave-particle model* by itself is fully able to account for the *cosmological redshift* of the Hubble diagram *without any need for the grossly more complicated hypothesis of expanding space.*

II. EXPERIMENTAL CONFIRMATION

Experimental confirmation of the *wave-particle model* for the *cosmological redshift* has now been obtained from *redshift* data received from Wernher von Braun's Pioneer 6 spacecraft. Accardi's Fig. 1, which is reproduced here, exhibits the redshift data received from Pioneer 6 as the spacecraft passed behind the Sun, the spectrally pure carrier wave (2292MHz) signal from the spacecraft passing through the sea of *free electrons* in the Sun's surrounding plasma. Distance d is the *distance* of the Pioneer 6 radio signal from the Sun in units of solar radii. In the graph, Accardi plots a prediction of *redshift* as permitted by *quantum electrodynamic* theory. The *wave-particle* prediction for *redshift*, which is based on a specific and well-defined mechanism, is here superimposed (in bold) on Accardi's Fig. 1.

Redshifts are calculated using the known *electron densities* in the plasma surrounding the Sun[3], and Eq. (5) (Ref. 1) which states that for small *redshifts* $z = Cnx$, or $z = \int Cn(x)dx$, where z is *redshift*, C is the *Thomson scattering cross section* for *electrons*, n(x) is the *electron density* of the plasma at each point, and x is the *distance* traveled through the plasma by the signal from the Pioneer 6 spacecraft.

[3] Guhathakurta, M. & Sittler Jr, E. C. (1999). *Astrophys. J.*, 523, 812.

The graph shows that the *wave-particle model* conforms reasonably to the experimental data.

III. CONCLUSIONS

The *wave-particle* explanation for the *cosmological red shift* with *no expansion of space* has now received experimental support. The unwieldy hypothesis of expanding space as an explanation for the *cosmological redshift* is now superfluous and unnecessary. Freed from the constraints imposed by *expanding universe* models, it finally becomes possible to understand phenomena such as *quasars* and *gamma ray* bursts, as well as *cosmic rays*, as being attributable to *matter-antimatter annihilation* in a universe composed of equal amounts of *matter* and *antimatter*[4].

[4] Mamas, D. L. (2011). *Phys. Essays*, 24, 475.

In order to progress into the new century, modern astronomy needs to be completely and finally freed of the unnecessary constraints imposed by expanding space models which may now be discarded.